Heart of
Dankness

Also by
Mark Haskell Smith

Moist

Delicious

Salty

Baked

Heart of Dankness

Underground Botanists, Outlaw Farmers, and the Race for the Cannabis Cup

Mark Haskell Smith

Broadway Paperbacks · New York

Published in the United States by Broadway Paperbacks, an imprint of the
Crown Publishing Group, a division of Random House, Inc., New York.

www.crownpublishing.com

Broadway Paperbacks and its logo, a letter B bisected on the diagonal, are
trademarks of Random House, Inc.

Cultivating and consuming marijuana for medical use is legal in the state of
California and, at the same time, illegal in the United States under federal
law. Because of this, the author has changed names and identifying details
to protect the privacy of some of the individuals in this book.

Library of Congress Cataloging-in-Publication Data

Smith, Mark Haskell.
  Heart of dankness : underground botanists, outlaw farmers, and the
race for the Cannabis Cup / Mark Haskell Smith.—1st ed.
      p.   cm.
Includes bibliographical references.
1. Marijuana—Therapeutic use—United States.   2. Smith, Mark
Haskell—Travel.   3. Marijuana industry.   4. Marijuana—Law and
legislation.   5. Cannabis—Competitions—Netherlands—Amsterdam.
I. Title.

RM666.C266S62 2012
615.3'2345—dc23

                                             2011025230

ISBN 978-0-307-72054-2
eISBN 978-0-307-72055-9

Printed in the United States of America

Book design: Ralph Fowler / rlfdesign
Cover design: Steve Attardo
Cover photographs: © Sandra Reccanello / SOPA / Corbis (Damrak Canal);
Hal Bergman / Vetta / Getty Images (diner)

10 9 8 7 6 5 4 3 2

First Edition

For Mr. Jones

UNDERGROUND BOTANISTS

Aaron, *DNA Genetics,* Amsterdam
Don, *DNA Genetics,* Amsterdam
Franco, *Green House Seeds,* Amsterdam
Arjan, *Green House Seeds,* Amsterdam
Reeferman, *Reeferman Seeds,* Moose Jaw, Saskatchewan, Canada
Swerve, *Cali Connection,* San Fernando Valley, California

OUTLAW FARMERS

Crockett, Sierra Nevada mountains
The Guru
Slim
Jerry
E

TRIMMERS

Red
Chuva
Cletus "The Dingo" McClusky

ACTIVISTS

Debby Goldsberry, Oakland
Richard Lee, Oaksterdam University, Oakland

CURATORS

Jon Foster, *Grey Area,* Amsterdam
Michael Backes, *Cornerstone Research Collective,* Los Angeles
Eli Scislowicz, Berkeley Patients Group, Berkeley

Heart of
Dankness

Chapter One

Paradiso

The *Paradiso* shifted in the water and cut its engine as it swung into the Brouwersgracht canal. The windows and roof of the boat were glass, making it look like a floating greenhouse, and tourists sat planted in rows, like some kind of sentient flora, taking in the sights of Amsterdam on a crisp November afternoon.

The static-spiked voice of a prerecorded tour guide rumbled from speakers inside the boat, a multilingual travelogue chock-full of facts and delivered with all the enthusiasm of an airport security announcement. The passengers followed the narrator, craning their necks in unison, perfectly synchronized sightseers.

I sat on a bench overlooking the water and watched the passengers blink up at the architecture behind me. The buildings alongside the canal slouched, leaning against each other for support like drunk friends, posing for the tourists who raised their cell phones and digital cameras, concentrating their gazes on the tiny images in their hands, oblivious to the life-sized versions that stood in front of them. From their awestruck and excited expressions I could tell that this was the site of something historical. Something momentous must have happened here.

The *Paradiso* revved its engine and, with a sputter and burbling surge, continued down the canal. The sound of the boat's motor faded and the sound of Amsterdam—the chime of a bicycle's bell, the squeal of children playing, the clang and rumble of a tram—replaced it.

The *Paradiso*'s propellor had churned the Brouwersgracht, leaving a wake that bounced off the sides of the canal and caused the water to ripple and refract, the surface reflecting the fading afternoon sun, turning slate, then blue, then violet, and then a color I'd never seen before, a color you can't find on any color wheel. It flickered and flashed along the canal like a special effect from a 3-D movie. It was breathtaking—the kind of blue Pantone would kill for.

It was my first trip to Amsterdam. I'd come to check out the annual *High Times* Cannabis Cup, to do some last minute fact-checking on a project, and to chronicle the experience for the *Los Angeles Times*. I wanted to experience the coffeeshops, to see what legal cannabis consumption might look like, and maybe go to the Van Gogh Museum and eat some pickled herring or *pannenkoeken*. But the real reason—the reason behind the other reasons—was to sample what was purported to be the best marijuana in the world.

I was not disappointed.

It was nothing like the Kansas dirt weed I'd smoked in high school. Back then we'd pile into Mark Farmer's bedroom after class, light up a doobie, and choke on its harsh smoke, getting as high from oxygen deprivation as from the scarce tetrahydrocannabinol in the crumbly leaves. We'd pass the shoddily constructed joint around—as seeds exploded inches from our faces—while we discussed the important teenage topics of the day: girls, motorcycles, girls and their breasts, electric guitars, what certain girls would look like naked, and the astonishing ass-kicking abilities of Bruce Lee.

The all-important incense would be lit—to cover the smell of the dirt weed—while we settled into a languid stupor, the faint taste of strawberry-flavored rolling papers on our lips, and let the power chords of some stoner epic like the Who's *Quadrophenia* or the insistent cowbell and corpulent lead guitar of Leslie West on Mountain's *Nantucket Sleighride* wash over us as we sipped Dr Pepper and stared at black-light posters of topless faeries posing in a psychedelic garden and Spanish galleons at sea.

A half hour before I watched the *Paradiso* putter up the canal, I'd sat in a sleek, modern coffeeshop called De Dampkring on Haarlemmerstraat in Amsterdam's city center. Seven German skinheads sat across from me, lined up along a banquette as if they were about to watch a soccer match. They were taking turns doing rips—inhaling hits of marijuana—from a large glass bong. The pot had energized them; they bounced off each other like hipster Oompa-Loompas, all playful punches and fake kung fu, brimming with testosterone, their faces plastered with oversized grins. They were distinctly nonthreatening skinheads, too goofy to be soccer hooligans or Nazi sympathizers, the shaved heads more a fashion statement—the look that goes with a hoodie sweatshirt, jeans, and logo Ts.

Next to me a stylish French couple snuggled and shared a nugget of hashish while two giggling Japanese girls in Tesla-high platforms and short skirts tromped downstairs to the bar for another round of bright orange Fantas.

The coffeeshop was festive, relaxed, like it was some cool person's private party and everyone was a VIP. As the good vibrations rolled around me, I sat at my table, sipped a cappuccino, and struggled to roll a joint. Intellectually I understood the technique of rolling. I ground the weed to a uniform size; I creased the paper and filled it with the recommended amount. I even took the little piece of thin cardboard and rolled it into a tight circle to act as the filter end. But when it came to the gentle massage, the all-important caressing of the paper in my fingers, rolling it back and forth to even out the bud and make a nice tight cylinder, I might as well have had lobster claws for hands.

The skinheads noticed my struggles and offered me the bong. I politely declined. Earlier, when I had gone down to the bar, I'd watched the waitress scrubbing out the clear glass tube with a toilet brush.

Eventually I crafted a ramshackle fatty and got it lit. I was smoking a cross between a Congolese sativa and a strain called Super Silver Haze, a hybrid named John Sinclair, after the former manager of Detroit punk rock pioneers the MC5. Smelling of sweet pine and tropical florals, this strain was De Dampkring's entry to the 2009 *High Times* Cannabis Cup, the premier marijuana-tasting competition on the planet.

What the Concours Mondial de Bruxelles is for wine, the Cannabis Cup is for weed: a blind tasting to determine the best of the best. Sponsored by *High Times* magazine, it's the Super Bowl of cannabis, the Mardi Gras of marijuana, the stoner equivalent of the Olympic Games for the botanists, growers, seed companies, and coffeeshops who compete. It's a harvest festival, a weeklong bacchanal, a trade show, and a deadly serious competition all rolled into one. Cannabis connoisseurs, marijuana industry professionals, and people who just like to party, all descend on Amsterdam to sample the entries. Most of them come from the United States, Canada, and various parts of Europe, but I've met people from as far away as Africa, Japan, and Brazil at the Cup. It is, truly, a global event.

While just being entered in the Cannabis Cup is a big deal for a lot of competitors, emerging victorious is even more valuable—just like winning a gold medal is for a vintage of wine. The competition ultimately determines the market value of the seeds and, make no mistake, for the botanists and seed companies who create these new strains, a Cup winner is potentially worth tens of millions of dollars.

My attempt at a joint collapsed like a badly wrapped burrito, but I had smoked enough of it. An anime-flavored techno track began galumphing out of the sound system—a tuba riff played under a glockenspiel melody with vibraslap punctuation—and the seven German skinheads became even more animated, clowning with one another in a kind of pixilated slow motion. I smiled. I wasn't stoned—I didn't feel like my ass was vacuum-sealed to my seat. I felt uplifted. I was energized, optimistic. In fact, I felt like taking a stroll.

Which is how I ended up alongside the Brouwersgracht canal, watching extraterrestrial colors spark and flare across the water.

As the *Paradiso*'s wake settled and the light show on the canal began to fade, a cartoon thought bubble popped up over my head, another special effect, as if epiphanies came with a bonus track. I looked up to see what I was thinking.

Set in Comic Sans and floating above me in the crisp November air were the words "This shit is dank!"

I had no idea what that meant.

Chapter Two

Superseded
by Damp

Back home in Los Angeles, I hauled out my massive *Webster's Encyclopedic Unabridged Dictionary of the English Language* and looked up the word "dank." It's defined as "unpleasantly moist or humid; damp." The example given is "a dank cellar or dungeon," which is pretty much what I always thought the word meant.

But dank wasn't always dank. In Rev. Walter William Skeat's *Etymological Dictionary of the English Language,* first published in 1882, he attributes the word to Scandinavia, primarily Swedish. There is nothing unpleasant or negative about these original definitions of the word; "dank" just means "moist or damp," a by-product of morning dew, and Skeat quotes several passages from early lyric poetry—"danketh the dew," and so forth—to prove his point.

In Swedish, "dank" can also indicate a "moist place," and that definition, at the very least, has a kind of naughty appeal. Or perhaps more logically—in a water pipe-y kind of way—it's derived from the Old High German "dampf" meaning "vapor."

The modern, Internet-friendly *Online Etymological Dictionary* compiled by Douglas Harper says that "dank" is from the fourteenth century, and is "obsolete, meaning 'to moisten,' used of mists, dews, etc. Now largely superseded by damp."

Superseded by damp? In what kind of world do we prefer the innocuous and banal word "damp" over the fecund and somewhat spiky word "dank"? If someone said their baby had a damp diaper, you wouldn't think much of it. Of course babies have damp diapers. But if they informed you that their child's diaper was *dank*, well, that's something entirely different. A diaper like that might be dangerous. It might require you to take precautions.

The fact that "dank" has been superseded by "damp"—the least sexy word I can think of—is like having your chocolate ice cream replaced by water-flavored air. Is that the kind of culture we've become? We're so boring that "dank" has been superseded by "damp"?

Fortunately "dank" didn't die. The original meaning mutated, and "dank" became a zombie word resurrected by a subculture, an undead word that can be used as a noun, a verb, or an adjective. Zombie dank roams the lexicon, hungry for brains.

The slang dictionary from the University of Oregon's Department of Linguistics credits snowboarders for giving the word a new lease on life. The adventurous young men and women who strap their feet to a piece of wood and career down icy mountains and leap off snowy berms to gain "amplitude" while they spin upside down and backward in the air created a new definition for dank. For them it means "strands of marijuana which has a very strong smell, and usually the pot itself is very tasty and potent."

Further exploration of slang and "urban" dictionaries available on the Internet reveal that "dank" now refers to anything of high quality or excellence and is sometimes used in place of the driven-into-meaninglessness cliché "awesome."

You have to admit that "dank" has some depth, a certain gravitas. It's about dark, moist places: caves and caverns, dungeons and cellars. It connotes strong, pungent smells and "vapors." But it also means something of superior quality, something that's awe inspiring. If you ask me, "dank" is a word resonant with a dark and kinky sexuality. It's like the Marquis de Sade of words.

But if dank marijuana was marijuana of the highest grade, what did that actually mean? The Cannabis Cup entry I tasted in Amsterdam wasn't the strongest pot I'd ever smoked. I've had bong hits that have left me poleaxed, like a post-lobotomy patient drooling on the sofa, which was an experience I didn't find particularly pleasant. What I tasted in the De Dampkring coffeeshop was different. It had a more nuanced quality, like you didn't know you were high until you noticed something that made you realize you were high, and then you were really, really high. And that was more than just a pleasant experience—it was a revelation.

I tried to find cannabis with similar qualities in Los Angeles, but despite their outlaw charm, black market dealers didn't have the breadth of selection I'd found in Amsterdam. The pot dealers I knew could offer up only variations of a potent strain that had originated in Afghanistan, Pakistan, and Northern India called Kush: they had Diamond Kush, Purple Kush, OG Kush, Hindu Kush, Master Kush, all kinds of Kush. These were all fine forms of knockout pot, but they didn't compare to John Sinclair.

I realized that to find something really dank, some connoisseur-quality cannabis, I needed to be able to go to a medical marijuana dispensary. And to do that, I needed to get legal.

Chapter Three

A Note from My Doctor

It was only a quarter past six on a Monday evening and the building was locked. I had expected the bank on the first floor to be closed, but the entire building? I'd heard Glendale was a sleepy suburb, but this seemed extreme.

I knocked on the door and heard the sound echoing through the marble lobby. No one answered. I pressed my nose to the glass and binoculared my hands so I could peek inside. A phone line blinked, unanswered, on the console of an empty security desk. A large ficus stood unmoving, silently converting carbon dioxide to oxygen, while a California state flag hung forlornly in the corner, a decorative afterthought.

I stepped back into the parking lot and looked up at the building. It was a modern structure, eight or nine stories high, the kind of building that's clad in black glass to look imposing and serious like an architectural Darth Vader. The building was dressed to impress.

There were a few lights still burning in some of the offices— perhaps even the office where I had an appointment.

I was considering my options—make a call, go home, sit down on the curb and wait—when a lean-looking dude in jeans and a black T-shirt opened the front door. He was wearing cowboy boots.

"You here for an evaluation?"

I nodded. He grinned.

"Take the elevator to six."

I'd been reluctant to get my medical marijuana card for a couple of reasons. One was political. I liked claiming the moral high ground when writing letters to the mayor, city council members, and the city attorney, who were trying to shut down the medical marijuana dispensaries in Los Angeles. I could honestly tell them that, although I myself was not a medical marijuana user, I strongly supported the rights of patients to freely and safely access their medicine. I guess I thought my letters might make more of an impact that way.

The other reason was that I was able to buy perfectly good pot from my local black market dealer, but that was before I went to Amsterdam and got a lungful of dank.

I had seen the menus of some of the local dispensaries and I knew that they were offering—or claiming to offer—some serious connoisseur-grade cannabis. The kind you might find in Amsterdam. Maybe even better.

The elevator doors slid open and a large laser-printed sticker—apparently stuck on the wall by someone unconcerned with hanging signage straight or at eye level—announced that I had arrived at my destination: the MMEC, or Medical Marijuana Evaluation Center.

In 1996, the Health and Safety Code of the State of California was amended. Section 11362.5, known as the Compassionate Use Act of 1996, effectively legalized the use and cultivation of marijuana for patients with a doctor's recommendation. It had been on the ballot as Proposition 215, and a majority of Californians had approved it. That act was further strengthened in 2003, when Governor Gray Davis signed Senate Bill 420, formally legalizing the right to consume and cultivate marijuana for medical use.

Californians are an entrepreneurial people. They came to Sutter's Mill looking for gold, they went to Hollywood hoping to be discovered, and they turned the Internet into the dot-com boom. If there's a way to make a buck out of nothing, to create an economy from thin air, Californians will be there. It's common in my neighborhood for enterprising chefs to set up an alfresco restaurant next to a liquor store or a florist's shop. They unfold a couple of card tables, line them with bowls of homemade salsa, fresh limes, and cilantro, and fire up the grill. If you don't mind eating standing up, you can get some of the best *carne asada* and *al pastor* tacos in the city. More sophisticated operators buy food trucks and leapfrog around town offering Mexican tacos, Korean tacos, Vietnamese tacos, Chinese tacos, ice cream sandwiches, dim sum, and whatever else people can eat quickly from a standing position.

Residents of the Golden State clearly understand the laws of supply and demand, so it didn't take long for doctors to open practices specializing in recommendations for medical marijuana, and for marijuana dispensaries to begin springing up to service those recommendations. In 2010, with more than eight hundred medical marijuana dispensaries operating in Los Angeles, the city attorney decided he had to act to control the "Wild West" atmosphere and attempted to shut more than half of them down. The city council backed this effort by rewriting the guidelines for dispensaries. This worked only temporarily. The dispensaries fought back. A judge ruled the council's actions were unconstitutional, and the dispensaries, for the most part, were back in business a few months later.

There are people who want to shut down the taco stands, too. And that's something I just don't understand.

I was greeted by a stocky young man as I entered the office. He was friendly and I immediately complimented him on his hair, which had been cropped into a flattop and dyed school bus yellow.

"Yeah," he said, "I like that old Orange County punk."

He smiled as he handed me a clipboard and a stack of forms. "You can fill these out in the waiting room."

He pointed to another area where several people—a couple of chubby Latinas in their twenties and a guy with a muscular build and a sunburned face—sat hunched over their clipboards under dim fluorescent lights.

I entered the waiting room and stopped. The space was the size of a basketball court. It was massive. And empty. It looked like a crime scene from the economic downturn. The carpet was scarred with ghost traces of cubicles, a monolithic warren of cubbyholes where workers had once clacked on keyboards, processed paperwork, and answered phones.

Along the far wall were offices with windows, where supervisors once supervised; the opposite wall was bare—the inspirational "Hang in there, Baby!" posters had been torn down—with only a few forlorn and abandoned filing cabinets breaking up the space.

The air was dusty; the room probably hadn't been vacuumed since the office had been shuttered and the fresh casualties of capitalism had packed their personal belongings in cardboard boxes—along with a company stapler or two—and shuffled out the door.

At one end was a large conference room, the kind of executive rah-rah lounge where PowerPoint presentations and team-building exercises had once flourished. The table and chairs were still there, poised and hopeful for the economy to revive.

The waiting area itself consisted of five plastic chairs and a rack of crinkly water-damaged magazines huddled together in a corner.

The evaluation center was essentially mobile: a computer, a printer, and a doctor; it could be set up anywhere. It reminded me of a hermit crab, an opportunistic crustacean that moves into another creature's abandoned shell and sets up shop.

I tapped my pen against the clipboard and heard an echo resonate from the far end of the room. I began filling out the forms and was immediately confronted by a straightforward question that required a complex answer. They wanted to know about my history with drugs.

One of the main criticisms of marijuana foisted upon the world by nicotine-addicted "drug czars," twelve-step sober presidents, and members of the law enforcement industry is that marijuana is a "gateway drug." The idea is that one puff of weed will lead to experimentation with stronger, more dangerous drugs, which ultimately ends with heroin addiction, prostitution, and a violent death in a squalid alley.

For me, Pabst Blue Ribbon was the gateway drug. It led to early teen experimentation with wine coolers and Canadian whiskey pilfered from our parents' liquor cabinets, often mixed with whatever was at hand in some kind of harebrained attempt at a cocktail. I still remember the taste of backyard punch made from a pint of Wild Turkey blended with cups of banana liqueur, blue curacao, and Jose Cuervo tequila diluted with a couple of lukewarm cans of Diet Pepsi. While most of us just reeled and unleashed streams of sweet vomit into the bushes, the strength of this concoction compelled one of my high school pals to go home and take a shit in his parents' bathtub.

But the gates were open and it wasn't too long before one of my friends got some pot from his older brother and we scurried down to the creek that ran between our suburban homes to smoke it. Nothing happened that first time, but we felt cool doing it, so we did it again. And again. And then, something happened. I got stoned.

Over the next ten years I took many drugs, many times, developing a real fondness for psychedelic mushrooms and, later, sauvignon blanc.

After college it was not uncommon to find my friends and me lounge-lizarding in our thrift store suits, drinking martinis liberally spiked with ecstasy at the Four Seasons Hotel in downtown Seattle or sweatily snorting narcotics in the bathrooms of nightclubs.

At some point cocaine began to demand everyone's full attention and, party pooper that it is, ruined everything.

I stopped taking drugs and underwent what the author Jerry Stahl calls "geographical rehab." I moved from the Pacific Northwest to Los Angeles to attend graduate school. For fifteen years or so I didn't take any drugs of any kind until I began to attend dinner parties with my South American friends. At some point, usually after the flan was served, a joint would be passed and I would take a couple of hits in the hopes that it would improve my Spanish. For a long time, I associated smoking pot with eating flan.

I was unsure if the medical marijuana evaluation form really required this kind of detailed information so I wrote "N/A" in the space provided.

The next page was a comprehensive list of the various medical conditions that cannabis can help alleviate or cure. There were hundreds. From patients undergoing chemotherapy and radiation treatments, to movement disorders, anxiety, arthritis, AIDS, migraines, chronic pain, lupus, fibromyalgia, Osgood-Schlatter disease, sleep apnea, diabetes, multiple sclerosis, motion sickness, chronic fatigue syndrome, and the list goes on and on.

I wasn't suffering from any of these conditions. I do, however, have intermittent insomnia and a rash on my face that becomes Gorbachev crimson when I'm stressed out. Because cannabis is one of the best stress relievers available, getting a doctor's recommendation seemed like a no-brainer.

The doctor stepped into the waiting room and invited me to follow him to the corner office, where the trappings of a former CEO—or at the very least a regional district manager—still remained. The doctor settled into a deluxe high-backed leather chair and spread my paperwork across a sweeping biomorphic-shaped desk.

He was older, with kindly Marcus Welby eyes behind thick lenses, and he spoke with a soft New England accent. He gave off a whiff of world weariness, like a retired cop or a college professor who'd been denied tenure. In fact, the only thing that indicated he was a medical professional was the lab coat he wore. Bleached a blinding white and crisply starched, the coat was authoritative and reassuring. It somehow did not strike me as peculiar that, embroidered above the pocket, just under his name, was the word "Gynecology."

I wondered what his story was. How did a gynecologist end up in an abandoned office dispensing medical marijuana recommendations? What tragedy did he suffer? Was he running from malpractice? Heartbreak? A late-blooming midlife crisis?

Given the fly-by-night charm of the enterprise, I expected my consultation with the gynecologist to be a kind of elaborate charade, a brief discussion of my ailments—the subtext being a somewhat cynical understanding that the result was a foregone conclusion—a determination that, indeed, I could benefit from medical cannabis, and I would walk out with a freshly laser printed recommendation.

But the gynecologist surprised me. He wasn't interested in

playing charades. He looked at my paperwork, then looked at me. He was skeptical.

"Your dermatologist said your skin condition was caused by stress?"

I nodded. She had, in fact. But he, apparently, didn't believe me.

"There is no evidence that I know of to indicate that stress is a relevant cause of this condition," he said. And he said it with some authority.

I wanted to reply that perhaps, because he was used to inspecting the genital areas and reproductive organs of women, he might not be exactly up to speed on skin conditions and their causes, but I didn't. I did, after all, want something from him.

"My dermatologist gave me a topical antibiotic."

He smiled.

"That's the correct course."

"But it comes back when I'm stressed and I don't want to take stronger antibiotics. She's big on antibiotics."

He scoffed at me, clearly annoyed.

"I'm sure she's *big* on them because they're effective. That's typically why we doctors are *big* on things. For example, if you came to me with a stress-related condition, I would probably be *big* on prescribing Prozac or some other antianxiety medication."

I didn't want to speculate on the chain of circumstances that might somehow lead me to visit a gynecologist for a rash on my face, but I also didn't want to irritate him further. He seemed genuinely annoyed that I was wasting his time with my feeble and transparent attempt to get a recommendation. As he sat behind the imposing desk glowering at me through his thick lenses, it suddenly occurred to me that this wasn't an elaborate charade after all. There was a very real chance that I might fail. It also occurred to me that he was waiting for me to say something, so I tried to rally with a reasonable-sounding argument.

"Why would I take a mind-altering chemical pharmaceutical when I can take a natural herb?" I asked.

This made him laugh. He looked at me again, really trying to see something, like he had gynecological X-ray vision. I think he was waiting for me to squirm, to blink, to give him some indication that I was a big fat phony. But I didn't. I sat still and met his gaze and affected a look of vague disinterest.

This peculiar stare down seemed like it might go on indefinitely and then, suddenly, he said, "I think cannabis can help with your insomnia."

And then he began signing the forms.

Chapter Four

Retail Weed

The same California laws that allow doctors to recommend marijuana for a variety of medical uses also allow cooperatives and collectives to sell pot to patients. The actual sale, whether it's classified as a donation to the co-op or a taxable item, seems to be open to interpretation, but the reality is that you can walk into a storefront dispensary and purchase cannabis without being subject to arrest or prosecution.

Frankly, it's a big relief.

According to a website called weedmaps.com, I have a dozen medical marijuana dispensaries within a two-mile radius of my house. Some of them are discreetly tucked into strip malls and office buildings without any signage, like speakeasies from the 1930s. They have names such as "Cornerstone Research Collective" or "Medical Caregivers" discreetly placed on small plaques by their doors. Others have names like "Emerald Bliss" or "Hyperion Healing" that make them appear to be day spas or health food supplement stores. Then there are a few that take a more aggressive, in-your-face approach. These are flashy storefronts with large green crosses blinking in bright neon with names such as "The Farmacy" or "Another World Chronics." Some of these places look like legit, corporate medical establishments, while other ones feature facades festooned with giant murals of hummingbirds and psychedelic flowers with typography borrowed from old Grateful Dead posters. There is, essentially, something for everyone.

My first experience with one of my local dispensaries was brief and, to be fair, it happened before I had my doctor's recommendation. But I was curious about dispensaries, how they operated, and what they looked like. Did they look like coffee-shops in Amsterdam?

The Eagle Rock Herbal Collective was a couple of blocks from my house, stuck between a liquor store and a Laundro-mat, which—if I had to sit in a Laundromat for a few hours and watch my whites, brights, and darks spin—seemed like a perfect location.

I opened the mirrored glass door and walked in.

Inside it was dark and smoky, with a mildewy smell like someone's basement rec room. A young Latino dude in an L.A. Dodgers replica jersey sat behind the desk, a large dog curled up at his feet. I realized it was the dog that smelled like mildew.

As I was about to introduce myself, a large man with a shaved head and a scraggly Zapata-style mustache, his gut thrust forward like the grill of a semi and framed by a pair of suspenders, ambled out of the back room and looked at me with what I can only describe as the stink eye. He was, apparently, the boss.

I extended my hand.

"I live just up the street and I'm working on a book about cannabis. I was wondering if I could talk to you for a minute about your operation."

The stink eye narrowed into a slit of angry suspicion.

"You got a recommendation?"

I shook my head. I hadn't visited the gynecologist yet.

"Then get the fuck out of here."

"Would you prefer to talk outside?"

"Get the fuck out of here." He sucked in some air, perhaps catching his breath from the effort he'd expended telling me to leave, before elaborating.

"Now."

The stink eye gave way to a mad dog glare and, with that, he stepped toward me in a distinctly unfriendly way. The moldy dog that had been curled up on the floor stood, stretched, and began to growl.

I backed out. I may be foolish, but I am not a fool.

Now, armed with my doctor's recommendation, I decided to sample a couple of other, hopefully more friendly, dispensaries in my neighborhood. I wasn't going back for more stink eye and mildewed pooch; instead I visited a highly regarded collective tucked into a strip mall at the end of my street called American Eagle Collective.

The collective was located between a chocolate shop and a Thai massage business. It was, in its own little way, the perfect intersection of hedonistic delights stuffed in a seedy looking strip mall. A discreet sign said "AEC," while a sticker on the door expanded that with "American Eagle." The front door was mirrored just like the Eagle Rock Herbal Collective, and I wondered if this was some kind of architectural touch typical of marijuana dispensaries.

I entered to find a cramped waiting area and a bulletproof window. The walls were mostly black, although I detected a kind of Star Wars theme. I'm not sure if that was intentional.

I slid my recommendation through a slot under the window in exchange for a clipboard with a few forms to fill out.

As I sat reading the various warnings and cautions, a steady stream of customers came in, showed their IDs, and were buzzed into the main dispensary area. It was two o'clock on a Tuesday afternoon and the place was bustling like it was the commuter rush at Grand Central.

I turned in my paperwork and was allowed to access the inner sanctum.

The funny thing is, once you're buzzed through the door, you step into a black metal cage. When the door behind you closes, the cage is opened. I wondered if they'd gotten this technology from a zoo. It might not be tough enough to deter serious criminals, but it would definitely keep a tiger from escaping.

I stepped out of the cage and into a dim room. The only real light was coming from the case that held large Mason jars filled with various strains of marijuana. It had all the charm of a tropical fish store minus the fish.

The budtenders were young Filipino kids who looked like they should have been studying for their midterms and not chucking buds into green plastic vials, yet despite the nonstop turnover and the customers stacking up behind me, they were all smiles.

"What kind of sativas do you have?"

The budtender reached under the counter and pulled out a jar filled with bright green clumps of marijuana.

"Green Crack. It's pretty popular."

He unscrewed the lid and I sniffed at the Green Crack. It's a terrible name for a strain. It's green, obviously, but why associate a benign herb with an addictive coca paste? Or maybe I've got it wrong. Maybe Green Crack conjures up an image of a determined little sprout reaching up toward the sun, breaking through a tiny fissure in the hard concrete of the modern world. Is Green Crack a Pixar movie or a product of urban blight?

Although I didn't think the name was any better than Green Crack, I decided to try a gram of Trainwreck. The budtender was unsure whether it was the original California Trainwreck or the strain developed by Green House Seeds in Holland but, because I was a new customer, he gave me a gift bag with rolling papers, blueberry-flavored pre-rolled cones, and a lighter. It was a thoughtful gift, but when I got home and looked at the Trainwreck in the daylight, I realized that it wasn't the best-looking bud and it had almost no scent. It definitely wasn't dank.

If the American Eagle Collective was like the busy trading floor of an exotic tropical fish exchange, then the Organic Healing Center was a serene and well-lit yoga studio, with a few nice touches, including a modern couch, warm hardwood floors, and soft pale green curtains. I wouldn't be surprised to learn that they'd hired an interior decorator to do the place up.

The dispensary room was large and bright, the well-organized display cases as clean as any I'd ever seen.

The budtenders could have worked as extras in a movie of the week about the Mexican mafia, but they were friendly and welcoming, and I was impressed by their knowledge and the fact that they took real care handling the product. Where the kids of the American Eagle Collective were handling buds with their fingers and flicking them into plastic vials, the budtenders here used tongs and tenderly placed the buds into little green containers with nicely printed labels. The color of the plants popped, and when I leaned over a jar for a closer look I was hit by a fresh and pungent scent.

I left with two grams. One was called Super Jack, a cross between a strain called Jack Herer—named after the hemp activist and author of the seminal marijuana masterpiece, *The Emperor Wears No Clothes: The Authoritative Historical Record of Cannabis and the Conspiracy Against Marijuana*—and Super Silver Haze, which won the Cannabis Cup in 1998 and 1999. I also took the budtender's advice and got a gram of Maui Wowee, a pure Hawaiian sativa.

I have to say that when I took the strains home and sampled them, I didn't find them to be anything extraordinary. You'd think with the kind of pedigree these strains had that they would be amazing, but for some reason they just didn't live up to my expectation. They were good, don't get me wrong, and the budtenders were really nice, but they didn't even begin to approach the level of weed I'd experienced in Holland.

I was invited to visit the Gourmet Green Room, a dispensary in West Los Angeles, to sample a strain they've developed called Zeta. The GGR has three different locations, one in Venice, one in San Diego, and this one, just off the 405 freeway. It's a dispensary that boasts more than one hundred strains of cannabis and is, according to the proprietor, the hot spot where young Hollywood goes to get their medicinal marijuana. I'd heard raves about Zeta, but I'd never tried it. That's because you can only get it at GGR, and even then you have to know someone. It's an off-menu, behind the velvet rope, VIP-type weed and is, supposedly, one of the dankest strains to come out of California in years.

I was met outside the building by an affable security guard who looked a lot like a pit bull. He checked my doctor's recommendation and then opened the door. I walked into a tiny foyer and passed my paperwork through a slot in yet another bulletproof window. I was starting to get used to this gauntlet of small rooms, bulletproof glass, and steel cages.

Once my bona fides had been vetted, the pit bull apologized for putting me through any inconvenience and I was let into the dispensary. The Gourmet Green Room struck me as kind of a misnomer. It wasn't particularly gourmet looking and it wasn't even green; it looked exactly like what it was—a converted industrial space. Even though I didn't go down any stairs the space had the look and feel of a basement. Maybe it was the linoleum floor and the diffuse lighting, or maybe it was the absence of any discernible character, ambiance, or charm. But once I looked behind the counter I understood. The room was empty so that customers could focus on the massive wall of various marijuana buds in large glass jars behind the budtender. It was impressive.

I met the owner, Mike, a compact and wiry guy who looked like he could just as easily have been the owner of a chain of pizzerias. He was outwardly friendly and easygoing, giving off a just-one-of-the-guys kind of vibe, but I could tell that underneath the firm handshake and jovial bonhomie was a shrewd operator. He pointed me to the lounge where Doug, the vapemaster, was holding sway.

"This is the last call. Next week we won't be allowed to medicate on site."

"Why not?"

Mike shrugged. "We're only nine hundred ninety-four feet away from a Chinese Baptist church. We need at least a thousand for the new ordinance."

The Los Angeles City Council, prodded by Carmen A. Trutanich, the headline-hungry city attorney, was attempting to crack down on the explosive growth of medical marijuana facilities by reinterpreting the rules they operate by. Unable or unwilling to tackle the problems of violent street gangs and serious crime, Trutanich and his office were taking the easy way out, stealing a page from Rudy Giuliani's playbook and trying to make the city more "livable" by putting the squeeze on soft targets and low-hanging fruit. Despite a fierce legal battle and several injunctions, in October 2011, Los Angeles County Superior Court judge Anthony J. Mohr ruled in favor of the city, and both the American Eagle Collective and Organic Healing were slated for closure. The fact that they were trying to close legitimate businesses in the midst of one of the worst recessions in Los Angeles history didn't seem to register with the city council or the mayor, who'd jumped on the anti-dispensary bandwagon.

I looked at Mike and shook my head.

"They're on the wrong side of history."

Mike's face lit up. "I like that. I like the way you think," he said.

I entered the lounge and noticed a large-screen HD television broadcasting the red carpet arrivals at the Oscars. Penelope Cruz, looking lovely in a deep red gown, posed for the paparazzi in front of gigantic statues of gilded alopecia sufferers, as other celebrities were herded past like prize pigs by their handlers.

Doug had been waiting. Lean and grizzledly handsome, he was a surprising choice for vaporizer operator of the GGR lounge. I would've expected a young stoner, but Doug was easily in his early sixties.

He stroked the short gray stubble on his chin and looked at me, barely suppressing a smile.

"You ready to try some Zeta?"

I was indeed.

Demi Moore strutted her cougar stroll on the red carpet dressed in a ruffled salmon-colored freak-out. The dress looked like a cake you'd order from an insane asylum.

As the Volcano vaporizer—which looks like a sophisticated version of every kid's fourth grade science fair project—warmed to 380 degrees Fahrenheit, Doug ground some of the Zeta in a small silver grinder. He was an intelligent and loquacious fellow, to say the least. In fact, he never stopped talking.

"I think marijuana is the next phase, the next step in human consciousness. You know? The secret's the sativa. It takes you up. Zoom. You soar, but you're in control. You're high but you're not stoned."

On the red carpet, someone I've never heard of grinned at the camera in a sharp-looking tuxedo. Then they cut to commercials.

Doug continued. "It knocks down the blocks, whatever it is that's blocking you, and you can be more creative. Don't you think? I mean, for me that's how it works."

I don't think I've ever met anyone as enthusiastic about anything as Doug was about cannabis. If the marijuana movement needs a head cheerleader to rally the team or someone to act as the bong-spinning drum major at the head of the parade, he's the man.

I'm not sure about creativity and cannabis. For me the jury's out. "I'm looking for dankness," I said.

"Ah, yeah." Doug handed me an alcohol swab to clean and sterilize the vaporizer mouthpiece. He shrugged. "Flu season."

I wiped down the little plastic piece and handed it back. Doug carefully packed the ground Zeta in the vaporizer chamber and affixed a plastic balloon on top. He put the chamber in the mouth of the volcano and flipped a switch. The balloon began to expand, filling with the mist of Zeta, something that I realize sounds like a role-playing computer game popular with fourteen-year-old boys.

"See if this is dank enough for you."

Doug handed me the balloon and I took a long inhale. I exhaled and got a blast of fresh sage and lemons. The flavor was fantastic. Another inhale and I could feel the scamper of THC rampaging through my nervous system.

Zeta delivered a soaring high that was lucid *and* euphoric. *Now this is what I'm talking about.* I could tell that Zeta was predominantly sativa, but I was curious what else was in the mix.

"What is it?"

Doug leaned back and exhaled a lungful of mist.

"I think it's Sage crossed with Blueberry and Trainwreck. But I don't know for sure."

The fact is, if he did know, he wouldn't say. Zeta is a topsecret strain developed by someone named Buddy, grown somewhere in California, and only available at the Gourmet Green Room.

On the television, George Clooney was working the red carpet. His smile had never looked shinier. Behind him, Sarah Jessica Parker grinned like a death skull, her bones looking like they were trying to poke through her skin. I almost asked Doug if we could turn the TV off, but then Woody Harrelson brought his swagger and a knowing twinkle to the proceedings. Woody's presence was reassuring. Perhaps he'd had a little Zeta before he got out of the limo.

Doug leaned forward and, lowering his voice to a near whisper, told me that Mike was thinking about entering Zeta in the Cannabis Cup. A win would be a big deal for the Gourmet Green Room franchise. It would put them on the map.

"What do you think? Think it could win?"

Helen Mirren put a hand on her hip and cocked her head as a million flashbulbs exploded around her. She looked lovely.

I took another hit off the balloon and tasted the sage and lemon again. Zeta made me want to eat roast chicken.

"Yeah." I nodded. "Yeah, I think it could."

I continued to visit various dispensaries, looking for a strain that might come close to what I'd had in Amsterdam, but Los Angeles, it seems, was in the midst of a Kush craze. Californians like their weed strong and sledgehammer heavy. It was the opposite of the euphoric uplift I'd gotten from the Zeta, or the herb in Amsterdam.

I tried a bunch of different Kush varietals. I sampled Diesel, NYC Diesel, and a strain with the great name Chem Dawg.

Besides Zeta, the only strain that even came close to the experience I was looking for—the only strain that I could say approached dankness—was one called Headband. Headband had taken third place at the 2009 Cannabis Cup and was allegedly a Dutch remix of a California strain: California genetics treated with an Amsterdam aesthetic.

And still, Headband wasn't quite it. I was pretty sure it wasn't "diggity dank"—an expression that the University of Oregon Department of Linguistics slangsters had said was used to indicate super-high-quality marijuana—but was it dank?

I didn't know. How did weed get dank? What happened that made some marijuana so different than the rest?

I decided that I needed to go back to the source, back to the home of the men and women who toil in the hidden greenhouses, grimy basements, and rooftop gardens of Amsterdam.

Holland, with its relative tolerance of cannabis consumption, has created the climate for a culture of growers and botanists to experiment with different types of cannabis, to seek out rare and exotic strains, and test-market the results in the coffeeshops. These are men and women whose passion is discovering and growing dank weed. There's a reason that the Cannabis Cup, the benchmark competition for the world's best weed, is held in Amsterdam. It was time to go back.

Chapter Five

The Grey Area

Although I've heard Amsterdam's Schiphol airport referred to as "shithole" by a number of people, it's one of my favorite airports in the world. There is nothing quite like stumbling off a ten-hour flight—which for me means I've been awake for almost twenty-four hours—and entering Schiphol's main terminal. It's like waking up and finding yourself in a steampunk wet dream with a heavy dose of *Blade Runner* chucked on top. By that I mean it's both modern and old-fashioned, with a layer of sci-fi rococo futuristic nonsense ladled on top. The bones of the airport look like an old steel Eiffel-designed train station, but one with a dozen layers of modern Tokyo signage-riot running amok. It makes you recalibrate your head, to adjust to the vastness of the architecture, the in-your-faceness of the multilingual information streaming at you, and the sheer volume of travelers going to all parts of the world.

That there's a subway station and a shopping mall, a library, and some sort of casino in the terminal just adds to the chaos and fun. The first time I arrived at Schiphol, it took me a half hour to find the line for passport control and customs.

When I told people I was going to Amsterdam, the first thing they asked was "Is your wife going with you?" When I said no—she has a job that ties her to Los Angeles—I got two distinctly different responses. One was "And she's okay with that?" The other, which came with a kind of twitchy wink of the eye, was "Awesome."

Amsterdam. The name alone seems to conjure up fantasies of debauchery and licentiousness—the city that offers guests the opportunity to wallow in a trough of hedonistic pleasure.

Amsterdam is a Bacchanalia, an orgy of sex, drugs, and stuff your mother warned you about.

Amsterdam was founded sometime in the thirteenth century. In a typical Dutch example of pragmatism over poetry, the city got its name when the Amstel River was dammed and the locals were called "the people who live near the Amstel Dam." Since then Amsterdammers have survived a few bouts of plague, the rise and fall of tulip mania, invasions by Napoléon and Hitler, and a misguided attempt at urban renewal.

I rented a basement apartment on a small side street called Utrechtsedwarsstraat just a hundred feet or so from a bustling shopping street called Utrechtsestraat. It is a beautiful part of the city. The streets are lined with stately seventeenth-century homes with wisteria vines climbing up them and pubs with big windows that face the canal. There was a small park nearby with beautiful trees and tulips sprouting in flower beds.

For me, this intersection could not have been more perfect. There was a pub on the corner called De Huyschkaemer that served a really good sandwich and fresh Grolsch beer. If I turned left I would find a chocolatier, a bookstore (with some English titles), and a very nice wine shop. If I crossed the street there was a tapas bar, a bakery, and a place called the Coffee Salon that made the best espresso I've ever tasted. In other words, all my favorite things—beer on tap, wine in bottles, books in English, and coffee in a cup—were just a thirty-second walk from my door.

My landlords were young, handsome, and charming: poster boys for gay marriage. In fact, I started to suspect that it was all a little too perfect, like there must be rats or mold or something unpleasant and moist in the newly redecorated basement flat, but the only disconcerting thing I found was a carton of something called "Vla" in the refrigerator.

A closer look and a quick browse in a Dutch-English dictionary revealed Vla to be vanilla custard.

I wondered if this might be a harbinger of things to come. Most Dutch people speak English, which gives you a sense of familiarity, and they look kind of like Americans, just a handsome, fitter, taller, and more cosmopolitan version. I imagine

Amsterdam is what Seattle might look like if everyone there was some sort of super-stylish metrosexual. Yet it's a foreign country, an alien culture, a society with customs and rituals that can seem totally bizarre to an outsider—like providing guests with a fresh carton of Vla.

I considered renting a bicycle. I wanted to have as close to a "real" Amsterdam experience as I could, and it seems like the entire population rides a bike. I'm not joking. There are literally thousands of dented, weather-worn, turn-of-the-last-century two-wheelers stacked up on every street and locked to every square inch of wrought iron railing in the city. Rush hour in Amsterdam is notably light on carbon emissions; car traffic is thin, while the bike paths are clogged with riders.

Like I said, I considered renting one, but I realized I didn't need one. My apartment was a pleasant twenty-minute stroll through the heart of the city to where most of the coffeeshops are clustered.

Although there are approximately two hundred coffeeshops in all shapes and sizes scattered across Amsterdam, Haarlemmerstraat is where the big names have parked their flagship storefronts. They're lined up, like a Las Vegas strip in miniature, all within a couple of blocks of one another. Barney's and Green House, easily the two biggest and best-known coffeeshops in the world, and the smaller but no less spectacular De Dampkring, have all built trendy and luxurious palaces for the consumption of cannabis on this stretch of street.

Barney's looks a lot like a bar in the lobby of a swank hotel, complete with computer screens at tables offering sightseeing tips, while Green House has a clear glass floor so you walk over an aquarium filled with decorative koi. De Dampkring is the techno choice, sleek and stripped down, high tech and minimalist with lots of brushed steel.

The idea of coffeeshops, places that sell cannabis and hashish over the counter, might seem unusual to most North Americans, like it's somehow shady, the kind of place where you might fall in with the wrong crowd. But when you enter one, they are, for the most part, similar to friendly cafés or restaurants anywhere in the world. Once you get past that initial Drug Abuse Resistance Education–inoculated uncertainty, coffeeshops are actually nice places to hang out in.

Before I left Los Angeles, I spent hours skimming back issues of *High Times, Skunk, Weed World,* and *Cannabis Culture* magazines not, as you might think, to ogle the glossy color photos of the various Miss *High Times* or 420 Girls—not that there's anything wrong with hot young women sucking on bongs or covering their breasts with pot leaves—but to get some kind of clue as to which coffeeshop in Amsterdam had the best reputation for quality. I canvassed people who'd attended the previous Cannabis Cup. I consulted strain developers and growers.

I heard a lot of names. In addition to the big three, there were Dutch Flowers, De Tweede Kamer, Green Place, Amnesia, and a few others, but one coffeeshop in particular kept being mentioned as a place for consistently superior cannabis: Grey Area, owned and operated by an American aficionado of all things cannabis named Jon Foster.

It wasn't easy to pin Jon down. He was busy, which I understood, and naturally skeptical of random guys who claim to be writing a book, which I also understood. But when I told him I was interested in dankness and what that meant, he became intrigued and agreed to meet me.

I strolled alongside the Singel canal until I came to Oude Leliestraat, the little street where Grey Area is located. It was almost noon and a group of young Americans, college students doing their semester abroad, were milling around, waiting for the door to open so they could begin their afternoon. A couple on their honeymoon stood in front of the window holding hands and making out, oblivious to everything but each other.

Jon arrived, right on time, riding his bicycle up to the front of the shop and dismounting with a practiced flourish. He's a good-looking guy, tall and thin, and on his bike he could easily pass for a local. Maybe that's because he's lived in Holland for almost fifteen years now. We shook hands and he opened the shop for business.

For a place that is renowned by cannabis connoisseurs as the mecca for the best weed in the world, Grey Area doesn't look the part. For starters, it's tiny. You couldn't park a car in the space—not an American car anyway. It has seating for maybe a dozen people and there is barely room behind the counter for the budtender to stand. Compared to the grand and luxurious decor at Green House and Barney's—upscale coffeeshops obviously owned by people with money to spend—Grey Area's furnishings are distinctly dinette set. And where Green House and Barney's have exotic art covering their walls, the walls of Grey Area are splattered—every square inch—with stickers. There are stickers for bands, punk shows, reggae festivals, motor oil, porn stars, restaurants, and random things that I was unable to decipher. Even a couple of old All Access passes to a Willie Nelson concert are slapped up on the wall. It sounds like it might cause your head to explode, but it's actually kind of cool. It's not as if the walls were covered with stickers when Grey Area first opened or it's some kind of *trompe l'oeil* gimmick painted by a professional; the stickers have grown on the walls organically, forming an archeological history of the customers who've visited.

It reminded me of the famous "Bubble Gum Alley" in San Luis Obispo, California. For more than forty years people have walked down an alley just off Higuera Street in the small central coast town and stuck their chewing gum on the brick wall. It's now a tsunami of gum, a museum of mastication, a minty monument to saliva and latex—and has somehow become a tourist attraction.

I like the no-frills vibe of Grey Area. It's similar to those fabulous old record stores like Bleecker Bob's in New York's West Village or Championship Vinyl, the record shop depicted in Nick Hornby's *High Fidelity.* There's a whack-a-doodle aesthetic at work there. It's an idiosyncratic temple to herb on a stylish European street and looks like a grad student's clubhouse.

If Green House and Barney's are the superstar rock bands of the cannabis world, the glittering palaces of ganja where most of the tourists go, then Grey Area is the indie rock favorite of the hipster cognoscenti.

Jon Foster thinks that's the point.

"We're a friendly, kinda lo-fi, living-room-type feeling place. And the combination of really dank weed with a dank atmosphere is really what brings it to another level."

There's that word again.

"What do you mean by 'dank'?"

Jon smiled and adjusted his chunky glasses. He pulled off his baseball cap, ran a hand across his head, and then reset his cap. He does that a lot when he's thinking, and he's a thoughtful guy. He reminded me of the smart, cool dude you knew in college who wasn't necessarily the life of the party but always knew where the parties were.

" 'Dank' is a word people use that's kinda undefined."

I wanted to interrupt him, to tell him about the University of Oregon's Linguistic Department, but he wasn't done. He was as curious about what "dank" means, on a practical and philosophical level, as I was.

"On a basic level the dankest weed, of course, comes from the strain type. A lot of people have told me you can only get so much out of a seed from, say, an unknown Mexican plant. You can grow that inside with a lot of care and it will still only be as good as, you know, what it is."

You hear people in the cannabis industry talk a lot about "strains." Botanically speaking, the term has no official significance: It's a catchall word that refers to the offspring of a plant that shares some characteristics with the parent plants. A more accurate term would be "varietal," which the wine industry uses to distinguish among different types of grapes. For example, sauvignon blanc, cabernet franc, merlot, Riesling, and zinfandel are all different grape varietals. The same can be said about cannabis and, believe me, the difference between a heavy indica and a light sativa is as profound as the difference between a fat red cabernet sauvignon and a tart, effervescent *vinho verde*.

Jon adjusted his baseball cap again. He'd given my question some serious thought and his answer surprised me.

"I feel like the dankest weed has a situational component to it. For example, the best weed I ever had was something someone grew outside and it was a gift and I was on holiday with my girlfriend, relaxing in a beautiful atmosphere, and all that really enhances the experience and really brought it a step up."

I looked around the diminutive coffeeshop and saw the college students relaxing around a table, drinking orange Fantas, and sharing a bong. The honeymooning couple were smoking a spliff and looking at a map of the city, trying to decide between visiting the Van Gogh Museum or the Sex Museum. Old-school jazz played softly in the background. The vibe was peaceful, serene, and timeless because the clock on the wall has been stuck at 4:20 for the past decade.

I realized that Jon was like a good chef, trying to control the "front of the house" experience for his customers, the situational factors that he believes contribute to dankness.

"That's why I like working in the shop. You see the people, you make a connection with them," he said. "And they're all happy because they're on vacation, experiencing something that maybe they've never experienced before. They're feeling positive and it makes me feel positive, too."

It's true that he's almost always smiling.

Grey Area is the only coffeeshop in Amsterdam that's owned and operated by an American, and I'm curious what strange twist of fate brought a nice young man from Rhode Island all the way to Holland.

"I was a drummer in a band called Love Box. We came over here to make a go of it, and I kind of stuck around."

Lots of musicians come to Europe to find audiences and play. Not many of them open coffeeshops. "Where did that impulse come from? I'm assuming they didn't offer Budtending 101 at Wesleyan."

"The concept was that this would be a vehicle for the music side—we'd have an income from this and could pursue the music. I'm still doing that, trying to balance everything. This is my livelihood so I can do my art."

I would argue that his coffeeshop is his true artistic expression, but perhaps it's more complicated than that.

What makes Grey Area different from the other coffeeshops in Amsterdam is not just the size of the place—it's the size of their menu. While most have dozens, if not hundreds, of cannabis varieties on their menu, Grey Area limits their offerings to eight or nine strains of marijuana and three or four kinds of hash. That means that Jon is incredibly discerning about what he sells in the shop. He must know what he's doing, as Grey Area consistently wins at the Cannabis Cup. That judgment isn't influenced by the situational elements; that's having some really dank weed.

Jon explained that Grey Area only sources and sells the highest quality cannabis available from select, small growers who grow the plants with what he describes as "love." He's not interested in what's popular or trendy in the weed world, and because he buys small lots, he can get odd and exotic strains, like a rare, and unnamed, equatorial sativa he had recently.

"We like to give new strains a run. We like to take risks, take the plunge, and see if it's something for the future. First we'll have our idea of what it is and then we put it to the public and see what they think. What helps us keep the weed on a really good level is to work with people who have an interest and a love for the product. It's artisanal cannabis."

Cannabis, artisanal or not, is technically illegal in the Netherlands. Not that you'd know it from walking down the street. The Netherlands Ministry of Foreign Affairs explains Dutch drug laws and the rationale behind them on its website.

They make a distinction between hard drugs, "substances which involve an unacceptable health risk, such as ecstasy, cocaine and heroin," and cannabis. Possession of cannabis for personal use—up to thirty grams—is a minor offense that is rarely, if ever, enforced.

One of the Dutch government's aims is to "separate the markets for hard drugs and cannabis." The government wants to protect casual cannabis users from "exposure to more harmful drugs." In other words, when I go to my local drug dealer to buy some weed, she usually has cocaine, LSD, mushrooms, and other substances for sale, but if I go to a coffeeshop, it's just cannabis and soft drinks. You can't even get a beer—the ultimate gateway drug—in a coffeeshop. It's a sensible and pragmatic approach that understands that people like to get high and that marijuana and hashish are not any different from alcohol.

Here's how the Ministry of Foreign Affairs says it: "The main aim of Dutch policy is to reduce both the demand for and supply of drugs, and to minimize any harm to users, the people with whom they associate, and the public in general." I especially like *minimize any harm to users.*

Although I'd prefer outright legalization of cannabis, if you have to have restrictions, what the Dutch call their "soft drug" policy seems to be a reasonable compromise. There is no rational reason why an adult should face fines and jail time for consuming a nontoxic plant in the privacy of his or her own home. The fact that, in the United States, there are people serving ten-year prison terms for growing marijuana plants in their backyards while Wall Street racketeers, who have defrauded millions of people and destroyed the global economy, walk free is a kind of bizarre hypocrisy that boggles my mind.

But if weed is technically illegal in Holland, what is, technically, a coffeeshop? Again, I turn to the Netherlands Ministry of Foreign Affairs.

"A coffee shop is an establishment where cannabis may be sold subject to certain strict conditions, but no alcoholic drinks may be sold or consumed. Although the sale of cannabis is an offense, coffee shops are not prosecuted provided they sell small quantities only and comply with the rules listed in C2."

Here are the rules:

- They may not sell more than five grams per person per day.

- They may not sell ecstasy or other hard drugs.

- They may not advertise drugs.

- They must ensure that there is no nuisance in their vicinity.

- They may not sell drugs to persons under eighteen or even allow them on the premises.

In addition to that, "coffee shops may stock up to 500 grams of cannabis without facing prosecution," which, as a busy coffeeshop will tell you, isn't that much weed. That's why most coffeeshops keep apartments or storage units nearby, and send runners for resupply when their inventory gets low.

According to Dutch government figures from 2008, there are 730 coffeeshops in the country, more than 200 of them in Amsterdam. It's hard to walk through the city center and not see a coffeeshop or two; they're right next to restaurants, bars, hair salons, and retail stores. If you don't see them, you can smell them—the sweet aroma of burning weed drifts in the air. Amsterdammers don't try to hide anything. Like the hookers in the red-light district, coffeeshops are part of the fabric of the city and a big part of the local economy. In 2008, coffeeshops in the Netherlands paid approximately 400 million euros in tax on gross sales of more than 2 billion euros. That's more money than the Dutch transportation system earns, and makes coffeeshops one of the biggest industries in the country. And that's not counting the four million tourists who come for the weed and stay in hotels, eat at restaurants, drink in bars, and visit museums, or the tens of millions of dollars generated from the sale of cannabis seeds.

Putting aside the humanitarian aspects of the Dutch policy, that's some serious financial incentive to keep the bongs bubbling.

But why do the Dutch have this policy and every other country is more like the United States?

"We Dutch don't like authority."

With that pronouncement, Joop Hazenberg, journalist, former Dutch government insider, founder of the political think tank Denktank Prospect, and author of the book *Change: How the Millennial Generation Will Conquer the Netherlands,* spread some liverwurst on a piece of brown bread.

I can't say I'm crazy about liverwurst. Liver is one of the few foods that I actively dislike; the smell induces a deep-seated revulsion that has more to do with my mother's bad cooking and my father's rage issues at the dinner table then the slimy brown stuff itself, but I decided to take a Dutch attitude and try to be tolerant.

Judging by the current standards of modern American journalism, Joop is CNN anchor handsome, with olive skin and dark eyes that flash with an intelligent, mischievous twinkle. "We're always finding little ways around the laws. Not because we're criminals, but because we don't like to be told what to do."

He mashed the brown bread down with the palm of his hand, and a shiny blob of liverwurst oozed out.

"Like the tobacco smoking ban in bars and cafés. Two-thirds of them don't care. People smoke. When the government tries to catch them, everyone sends SMS messages on their cell phones to warn that the inspectors are coming."

I put a slice of cheese on a piece of brown bread—an action that would become a recurring theme of my stay in Holland—and looked around. We were sitting at the communal dining table in the center of a large workspace shared by a group of freelance writers. Scattered throughout the floor of the building, which overlooked the Oudezijds Voorburgwal canal, were five or six desks with diverse levels of neatness. Some were clean, almost barren; others were buried under piles of paper, books, and knickknacks.

The floor was bare concrete, the walls unadorned except for an old movie poster from *Casablanca* tacked to the wall above the table and a large swatch of glittery purple fabric randomly draped along another. Bookshelves acted as room dividers and were stacked with magazines, old newspapers, and half-eaten boxes of cookies. There was an air of dilapidated faculty lounge about the place: It was part clubhouse and part creative commons, and I couldn't help wishing I had a desk in some corner.

Because Amsterdam boasts a large creative community, these kinds of arrangements are becoming more and more common, said Joop. They are called, and I think the irony is unintentional, "freelance offices."

I was having lunch with Joop because I was curious how Holland had become such a tolerant society. Here's a country that is a right-wing American's worst nightmare. Holland has socialized health care, gay marriage, legal prostitution, euthanasia, and coffeeshops where you can smoke marijuana. And, just to rub it in, its people are ranked among the happiest in the world.

"From the beginning we have always worked together. We had to. You tolerate a lot of differences when you're fighting to keep the sea from flooding your country."

I nodded and took a bite of my cheese sandwich. Netherlands means, literally, "low lands." It's a country that's mostly at or below sea level, the ocean held back by a series of dikes, embankments, and canals.

Joop held his liverwurst sandwich in the air—he was caught mid-thought. He put the sandwich back on the plate. "I think that's where it began. But then we have always been a country of successful traders and businessmen. The Portuguese Jews came here in the sixteenth century, the Huguenots after them. There are always communities of people coming. You can't be a successful trading nation without being open to the world and tolerant of other cultures."

It clearly annoyed him when I asked about the coffeeshops. His face contorted in irritation and he heaved a weary sigh. "Why is that the first thing people ask when they come here? 'Where are the coffeeshops?' Why? Don't they know that Amsterdam is much more than sex and cannabis?"

While it's a bit like someone from the Bahamas complaining that tourists only want to go to the beach and drink rum, this concern about the world's perception of the city is common for native Amsterdammers.

"You can't blame the tourists. Amsterdam has the best cannabis in the world," I told him.

This, apparently, was news to Joop.

"Really?"

I pointed out that millions of seeds and hundreds of thousands of kilos of *nederwiet*—Dutch-grown cannabis—are exported out of Holland every year. He considered that fact and, although he tried to hide it behind his sandwich, I detected a hint of pride.

We were interrupted by another writer, Minka, who joined us at the communal table. I watched as she carefully buttered her brown bread and dumped a pile of chocolate sprinkles called *muisjes*—little mice—on top. She spread the sprinkles carefully with a knife. I couldn't help staring. It's not often you see an adult eating a chocolate sprinkle sandwich for lunch.

Joop continued. "In 1996, we had a purple parliament."

"What?"

"The socialists were red, the liberal democrats blue. They formed a coalition government."

So it was not, as I had hoped, a government inspired by the pop star Prince.

"But the liberals here are not like liberals in the United States. Here they are more like libertarians. They don't want people telling them how to live."

Like the prostitutes in the red-light district, coffeeshops had been operating illegally, but without any serious enforcement, since the late '70s. There was never any formal regulation in place.

"The purple parliament licensed the coffeeshops and legalized gay marriage, euthanasia, prostitution—all the social reforms that we have now."

I watched Joop take a bite out of his liverwurst and then turned to see Minka chewing a big bite of chocolate sprinkle sandwich. It was one of the most iconoclastic lunches I'd ever attended. Maybe that's typically Dutch.

But it hasn't always been smooth sailing for Dutch tolerance. In fact, things are changing. According to Joop, a few recent events have shaken Dutch society and are making them reconsider how far tolerance can actually be extended.

After the attacks in New York City on September 11, 2001, a flamboyant, openly gay, and, for the Netherlands, right-wing politician named Pim Fortuyn declared that the purple parliament had made a mess of the country. He started his own political party—the eponymic Pim Fortuyn List—and ran a popular campaign to limit immigration and stop the multiculturalism that he felt was eroding traditional Dutch culture. In 2002, he was assassinated by a disgruntled animal-rights activist and vegan who felt that Fortuyn was picking on "weaker groups in society."

I looked at Joop and raised an eyebrow.

"A homicidal vegan?"

Joop shrugged. "We're talking about Holland."

The assassination of Pym Fortuyn was followed by the murder of filmmaker and writer Theo van Gogh in 2004. A Dutch-Moroccan man outraged by a film Van Gogh made depicting the deplorable treatment of women in Islamic countries shot him eight times—and then attempted to decapitate him with a knife—as he bicycled to work. For many Dutch citizens, this had as big an emotional impact as the 9/11 terrorist attacks had on the United States, and the traditionally tolerant Dutch public began to actively question their policy of openness.

These events gave rise to Geert Wilders, a right-wing politician who looks a lot like Siegfried or Roy, and is a virulent anti-Islamist who supports a moratorium on immigration from Muslim nations, a ban on the building of mosques in Holland, and a ban on traditional Muslim clothing like the burka. His logic is simple: Holland, he says, cannot afford to "tolerate the intolerant." Wilders flatly claims that if the Muslim population in Holland continues to grow and manages to force their intolerant views into Dutch political life, Amsterdam will no longer be the "gay capital of Europe."

I can't imagine an American politician, Democrat or Republican, seeking to ban immigrants to protect a culture of tolerance, gay rights, cannabis, and the other social freedoms that are taken for granted in Dutch society. In fact, FOX News television personality Bill O'Reilly famously claimed in 2008 that the Netherlands exemplified "extreme liberalism" and had become a "cesspool of corruption" caught in the "grips of anarchy."

Which is a load of horse shit. I don't see any evidence of anarchy or even intolerance. If anything, the Netherlands is, like a lot of societies, struggling to adapt to globalism and a changing world, only they're being a little more thoughtful about it.

Joop polished off his liverwurst sandwich and chased it with a glass of strawberry juice. Minka got up to go back to work, leaving a scattering of chocolate sprinkles on the table that I was tempted to eat. Joop wiped his lips with a napkin and got up to make some coffee.

"I don't see this as Islam versus the West or any kind of religious schism. That won't be the culture war."

"What is it then?"

Joop smiled as dark espresso bubbled out of the coffee maker.

"I think it's sovereignty versus cosmopolitanism."

I admit, I had to think about this for a minute. Cosmopolitanism is the belief that all humans share a similar underlying morality—we all want to be happy, to be free from suffering, and so forth—and that we belong to a community of humanity. It's a worldview that is inclusive, yet allows for individuals to have different opinions and lifestyles, and fits nicely with the "one love" ideology of the stoner set. Sovereignty, by contrast, is about excluding people who are "different." It's a kind of narrow-minded and arrogant tribalism.

I like that the Dutch do things their way. It's a society that values rebelliousness and tolerance. People are given license to be idiosyncratic and unique and live their lives the way they want to. It's a simple yet profound idea and it seems to work. It's a sharp contrast to the United States, which claims to be "the land of the free" and yet denies its citizens basic liberties like the right to marry who they want or the freedom to ingest nontoxic plants in the privacy of their own home. Yet not all Dutch people are for this quasi-legal soft drug policy. In fact, the country is split pretty evenly, with some Dutch politicians arguing for the closing of the coffeeshops, or limiting the sale of cannabis to Dutch citizens, while many, like Amsterdam's current mayor Job Cohen, are supportive of the industry.

Back at Grey Area, I asked Jon Foster about this and he shrugged.

"Every few years when there's an election, someone talks about shutting down the shops and then after the election that talk kinda goes away."

I explained that what struck me as funny—in an ironic way—was the part of the law that proclaims it is "an offense to produce, possess, sell, import or export drugs." In other words, coffeeshops can sell small amounts of cannabis but it's illegal for growers to grow the weed or importers to bring it in to the country. And yet Dutch-grown cannabis and Moroccan hash are the big sellers on every coffeeshop menu in Amsterdam.

Jon laughed. "Yeah. It's weird."

Coffeeshops exist in a wacky twist of legal logic, a kind of gray area—hence the name of Jon's shop.

I asked Jon how he selects the weed he carries. He's got a network of small, artisanal growers bringing their product into the shop, but it can't all be dank. What does he look for?

Jon lifted his cap and ran his hand across his head.

"We would look for crystals and if the plant looked like it had been grown well. Obviously if it looks pretty it's easier to sell than if it looks all scraggly. And then how it's dried. If the stems snap. And we'd want to make sure it wasn't too dry. A lot of people try to quick dry or flash dry the herb and it just breaks into dust."

He shook his head sadly and a look passed across his face that seemed to say it all. *Why can't people just be patient?* Then he continued.

"We smell for the aroma. We look for a clean, fruity smell. Sometimes there's a smell that's almost like fish because they use fish meal fertilizer. That's undesirable, for sure."

It sounded disgusting.

"And then we usually have to smoke it—which is the best test."

I'm not a cannabis connoisseur. In fact, I'm hopeless. I can't tell the difference between sativa and indica until after I've smoked it and I either get high—which I like—or get stoned and sleepy—which I don't like so much. I can't really tell if the bud has been cured properly, and I can't detect fertilizers or chemicals or fish meal in the taste of the smoke. Well, maybe I could taste the fish meal. I'd like to think I could.

Over the course of a few visits to Grey Area I sampled a number of the strains on the menu. The one thing that stuck out was that they were noticeably stronger than the strains I'd smoked at other coffeeshops. I told Jon how the Silver Bubble—a cross between Super Silver Haze and Bubblegum—kicked my ass. And by that I mean I had to go back to my apartment, lie down, and go to sleep.

He laughed. "I'm not surprised."

For me, that experience wasn't dank. It was like being abducted.

To make up for the Silver Bubble experience, Jon rolled me a joint of something called Grey Haze, which I took back to my apartment and smoked. This cannabis was strong—it gave me the "up" feeling I liked—and it had the unusual side effect of stopping time. I know that sounds like an outrageous claim, but it's true. I had a couple of hits off the joint. I didn't want to smoke too much because I'd agreed to meet someone for dinner and had about a half hour before I needed to leave. I checked the time: It was four thirty in the afternoon. With the Grey Haze rattling around in my brain and some time to kill, I began to clean up the apartment. I swept the floor, did the dishes in the sink, folded some laundry, got dressed, brushed my teeth, checked my emails, and drank a glass of mineral water. Then I lay down on the bed for what seemed like a half hour and, thinking I was late, jumped up and checked the clock. It was four thirty-four.

Grey Haze wasn't dank, but it did freak me out.

It seemed to me that dankness was more complicated than I first thought. If what Jon said was true, dankness begins with the genetics of the plant, then the care taken in growing and drying it is factored in, and finally there are the "situational elements." That seemed to me like a chain of unrelated incidents.

There are too many variables. You can't always control the where and how and who and when you smoke pot. It's not always with your lover in a friendly coffeeshop on a nice day in a charming European city. The growing and drying process can be controlled to a large degree, but that requires effort and constant vigilance. And you have to trust that the farmer didn't use herbicides or pesticides that could still be in the plant. Nobody wants to smoke bug spray. But can you control a plant's genetics? And if you can control it, can you breed a plant to be dank?

I realized that I needed to start at the beginning. I needed to take a class in basic botany.

Chapter Six

Botany 101

It's Koninginnedag, Queen's Day, and people all over the city are emptying out their closets and attics, piling their crap on the street in the hopes that someone will actually buy it. This is because the Dutch government has declared Queen's Day a *vrij-markt,* or "free market," so if someone is actually able to unload an old crib, that broken bicycle, or those golf clubs that have been rusting in the trunk of their Renault for the last seven years, the government won't charge a tax. The holiday is all about national unity, and nothing brings a nation together better than a good old-fashioned duty-free yard sale.

And, of course, beer.

And wearing the color orange, apparently.

I strolled up Utrechtsestraat to Rembrandtplein—a plaza I remember as the location of a restaurant where I suffered through a terrible Indonesian *rijsttafel*—an elaborate succession of a dozen or so disappointing and underspiced dishes slowly burning over small candles—and made my way up a street called Rokin. I passed the Royal Palace where a cluster of soldiers from the Royal Netherlands Army stood around looking spiffy in combat fatigues and berets while workers unloaded metal gates to create a protective barrier for the queen. This security was not unjustified.

In 2009, a recently fired security guard named Karst Tates drove his black Suzuki Swift at high speed through a crowd in an attempt to collide with an open-topped bus carrying the royal family. Paradoxically described as a "nice quiet fellow" and someone with "strong right-wing views," but not as a particularly good driver, he missed the bus and slammed into a concrete monument. Eight people, including Tates himself, died in the attack.

I left the main street and cut through some of the smaller side streets of the Amsterdam Centrum. These were narrow passageways lined with small bars, cramped restaurants, and, of course, coffeeshops where marijuana and hashish are consumed.

Outside of one coffeeshop, as the sweet, skunky smell of cannabis mixed with the scent of incense, causing my nose to go into a patchouli-overdrive sneeze, I passed a very stoned French girl standing next to a bicycle. She had a huge grin plastered on her face like some kind of maniac, and was waving excitedly to her friends, jumping up and down, as she rang the bike's bell over and over again.

It did have a pleasing tone.

I got turned around, an easy thing to do in a city where the streets run in long, looping U-shaped curves, and found myself in De Wallen, Amsterdam's famous red-light district. It's a major tourist attraction, like Disneyland, only with different kinds of rides.

Unlike pot smoking, which is technically illegal, paying someone for sex is a completely legitimate transaction in Amsterdam. I strolled through the alleyways lined with glass doors and windows where prostitutes stood, hands on hips, like lingerie-clad cuts in a butcher shop. The afternoon shift was just starting. While a few windows displayed women who looked to be from eastern Europe—not that I know, but they looked a little bit like Ivana Trump—the majority appeared to be from Africa, with squat and stocky bodies that looked powerful, as if they were recruited from the Ghanaian Olympic weightlifting team. Their flesh spilled out of what little they wore, like Sumo wrestlers stuffed into bikinis. They reminded me of the actor Yaphet Kotto, if you can imagine Yaphet Kotto in a pink pageboy wig and lacy lingerie.

Considering they were standing in the glass doors and windows to entice men inside, their expressions were not what I would call inviting. There was no come-hither bat of the eyes. They didn't purse their lips or beckon customers with a twitch of their fingers or a heave of their breasts. Instead they looked like office workers everywhere, stuck in a cubicle trying to make a buck—which is not to say they aren't good at their jobs. I imagine in a competitive business like whoring you can't just phone it in; you need to have skills.

De Wallen also boasts a number of sex shops, sex clubs, bars, and, of course, coffeeshops. At one porno video store a handwritten sign in the window advertised "Half Off All Animal DVDs."

I eventually made my way to a quiet street called Sint Nicolaasstraat, stepped over a massive rain puddle, and walked into the DNA Genetics store.

Rafaela, an attractive young woman with beautifully rendered tattoos randomly splashed across her body, sat behind the front counter and leaned over a white porcelain dish with an expression of absolute concentration on her face. She looked up and smiled at me before tucking a strand of dark brown hair behind her ear and turning back to her work. The dish contained a pile of tiny cannabis seeds, and it was Raf's job to separate the seeds from the bits of dried leaves and stems, inspecting them closely to make sure none were cracked or damaged, and then package them in small plastic bags. The work looked tedious, even migraine inducing, but she took to the task with good humor. It's just part of the behind-the-scenes glamour of working in the cannabis business.

She funneled some seeds into a plastic bag about the size of a shirt pocket. I was curious how many seeds were in the bag; I would've guessed five hundred, maybe a few more, but she surprised me.

"Maybe, I guess, seven thousand."

Aaron, the "A" of DNA Genetics, emerged from the back wearing a slime-spattered T-shirt and holding a garden hose. Aaron is one of the foremost strain developers in the world, famous in the field of cannabis genetics, but instead of the hipster scientist I was expecting, he looked like a plumber in the middle of snaking a badly clogged toilet. He apologized and said he was changing the water in the massive fish tank that takes up half the back wall of the store. As the water level in the tank dropped, a large mud-colored carplike fish looked out with an expression of aquatic anxiety.

Aaron's a good-looking dude with close-cropped hair and a lean and lanky physique that reflects his years of playing varsity soccer and baseball. He's friendly, with a quick smile and intelligent eyes, but there's a distinct South Central swagger, an authentic *don't fuck with me* undertone that comes from growing up in certain parts of Los Angeles. In other words: He has the potential to mad dog on a dime. It's easy to see why he's probably the only strain developer in the world who's a member of the Screen Actors Guild. In his early days in Hollywood he had a brief career portraying long-haired drug dealers on various TV shows. His cinematic specialty, he says, was "running from the police."

By contrast, Don, the "D" of DNA Genetics, exudes an affable, doughy charm. He's got a wry sense of humor and is quick to stroke his chin and chuckle with Dalai Lama–esque amusement at the world.

Aaron was about to go back to draining the fish tank when his cell phone rang. He apologized for the interruption; it was his wife calling to tell him she was bringing their newborn daughter to the shop. His face took on an expression of parental concern.

"Hey, babe. Watch out for the puddle in front of the store."

What's unlikely about the DNA story is how likely it is. It's the same rags-to-riches success story of any artist, inventor, or businessperson with a good idea and the chutzpah to turn that idea into reality. Only in Don and Aaron's version, the young entrepreneurs didn't come to America to make their fortune; they flipped the myth, abandoned the land of opportunity, and moved to Holland.

They met in the usual way: Aaron was a pot dealer and Don was one of his customers, but they soon discovered they had mutual interests. Both of them were frustrated by the low quality of the cannabis available for sale. "It was," they said, "complete schwag."

Like all innovators, they decided they could do it better. Or, at the very least, they could try to do it better. They were already experienced growers, so they started with what they knew and began experimenting, cross-pollinating various strains, growing out those seeds and seeing what the combinations created. They'd been saving seeds from various bags of weed they'd purchased over the years and they started trading seeds and cuts of cannabis with other growers, acquiring as many exotic varietals as they could. Working through trial and error—trusting their nose and instincts—they became self-taught botanists. It didn't take long for them to realize they were on to something special. Unfortunately, Aaron suspected the Drug Enforcement Agency and the Los Angeles Police Department might be on to something, too. Namely, him.

"So we came up with a plan. We moved over here to Amsterdam and I said, 'In two years we're gonna win the Cannabis Cup.'"

Although Aaron might not be quick to admit it, love had something to do with this decision, and not just love for the reefer.

He and Don supplemented their income by working as carpenters and painters. They were hired by a hip Melrose Avenue retailer to go to Brussels to help build a store. As Aaron tells it, "Long story short. He doesn't show up. We spend about four weeks waiting around in the freezing cold and rain and then I'm like, let's go to Amsterdam, bro."

Because they bought their return tickets for when they thought the job would be over, they had three months to kill before returning to Los Angeles. They got a job in Amsterdam refurbishing a youth hostel called the Flying Pig, and it was there that Aaron met Kim, a young Canadian who was working at the hostel.

When Don and Aaron eventually went back to California, Kim stayed, living in the city and working as a waitress. It didn't take long before Don and Aaron were back, only this time they had a plan.

True to Aaron's word, at the 2004 Cannabis Cup their strain L.A. Confidential won third place for best indica. A year later, in 2005, they burst into the cannabis world's consciousness with a rare double, taking first place for best sativa with Martian Mean Green and second place for best indica with a new, improved version of L.A. Confidential.

Botanically speaking, *Cannabis* is the genus of the family Cannabaceae, and the plant contains three distinct subspecies: *Cannabis sativa, Cannabis indica,* and *Cannabis ruderalis.* Each of the plants has unique traits: The indica is low and bushy and indigenous to the mountainous regions of India and Iran. It is a robust plant and has naturalized in such diverse parts of the world as Europe, North America, and Brazil. *Cannabis sativa* tends to be long and rangy, preferring hot, tropical zones, like Africa and Southeast Asia. *Cannabis ruderalis* is the red-headed stepchild of the group. Hardy and quick flowering, this low-growing plant has such modest levels of THC, the active ingredient in cannabis, that it hasn't been, historically, used for medicinal or recreation purposes, although various breeders have recently begun experimenting with hybrids using *Cannabis ruderalis.*

For decades *Cannabis indica* has been the species of preference for growers, as it needs less room to mature and is quicker to flower than a sativa.

L.A. Confidential, one of DNA Genetics' Cup-winning strains, is a serious indica, the kind of cannabis that gets you stoned, that knocks you on your ass and gives you a sensation often referred to as "couch lock." A lot of people enjoy indicas—they like becoming well-baked couch potatoes—but the current trend among serious connoisseurs is for a lively sativa, like DNA's Martian Mean Green, something uplifting and energizing.

But by far the most popular strains—and this is where underground botanists like Don and Aaron come in—are the hybrids, the mix between indica and sativa that offers the best of both worlds, a sativa-like high coupled with the muscle-relaxing effects, shorter flowering times, and bigger plant yields associated with indica. A classic DNA hybrid is Connie Chung, which is a cross between L.A. Confidential and a more sativa-leaning varietal called "G-13 Haze." It's one of the strongest strains they produce and, according to their website, guarantees "to make you 'Chinese.'"

In what way it makes you Chinese is unclear.

Aaron came out of the back room wearing a clean California Angels jersey—"I gotta represent Cali over here"—and waved me into the back office. Although there were desks and computers and lots of files and things generally associated with offices and business, it looked more like a stoner's dream lounge. There were stacks of T-shirts, hoodies, and hats; boxes filled with custom bongs and water pipes, and bookshelves containing their inventory of seeds. One wall was the back of the giant fish tank—the fish look much happier now—the other was filled with a spray-painted mural of a jolly green Martian.

I pointed to the mural and asked Aaron what it meant for them to win the Cannabis Cup.

An irrepressible smile bloomed on his face, but it wasn't arrogance or some kind of ego trip. His expression was a mix of bemusement and amazement, as if he still couldn't believe he was actually here, doing what he does.

"On a business level it's like having the first flight to outer space. It boosts your business, your reputation. It gets you known throughout the world."

"And on a personal level?"

"On a personal note?" He took a moment to scratch his goatee and consider the question. "On a personal note, it means that we did something right."

Aaron and Don have done a lot right. They've won more than forty major cups and awards from the cannabis industry since they opened for business in 2003 and DNA Genetics has become one of the top three or four seed companies in the world.

"How would you define 'dank'?"

"You mean a strain or a plant? Because sometimes, bro, a plant can be really dank. You know? You can just smell it. It doesn't matter what strain it is."

"I guess I mean, what strain do you, personally, think is your dankest strain?"

I admit it's a strange question. Aaron took his time to consider it. "Our Headband is dank. Chocolope is dank."

Then he leaned forward, his eyes flashing with a touch of that mad-dog passion, as he suddenly got serious. "L.A. Confidential, OG Kush, OG #18. There's a lot of dank around here, bro. I think in order for us to make seeds of it, we have to consider it dank."

"So dank is the minimum basic requirement?"

He shot me a look that said "Duh."

"There's a process. Is it good enough? If it's not good enough, it's gone, we're not gonna make seeds of it. So it has to be dank to start. Then there's above dank."

"Really? What's better than dank?"

Aaron doesn't hesitate. "Super dank. Holy grail dank."

Holy grail dank. I wonder what the Linguistics Department of the University of Oregon would make of that.

"I thought the holy grail is supposed to be unattainable."

"Yeah, but for me personally, I think there *is* a holy grail plant. It's called the Sleestak."

Sleestak, named after the green googly-eyed reptilian insect monsters from the '70s TV series *Land of the Lost,* is a relatively new sativa-indica hybrid developed by DNA.

"I only smoke dry-sift hashish or resin and not every plant makes good resin, but this is the best that I've found."

I wondered what was so great about it, and Aaron shifted into connoisseur mode.

"It's mostly sativa with an old-school flavor, kinda hazy on the exhale, great room smell, great uppity high. I mean, it can still knock you down, bro, but it's phenomenal. I love that strain. It's pretty much all I smoke."

"Is it the dankest of the dank?"

Aaron shrugged.

"I don't know, bro, but if I could grow rooms of Sleestak and make hash out of it, I'd be the happiest motherfucker on the planet."

I was curious, naturally, about this Sleestak dry-sift resin. Aaron took a small plastic cylinder out of his pocket. It was about the size of a roll of quarters. He opened a pocket knife and dipped the point into the cylinder, removing a tiny clump of straw-colored powder—it looked like sand—and tapped it into the bowl of a large bong. He slid a serious looking lighter toward me.

"Tell me what you think."

Grasping the miniature blowtorch, I crème brûléed the resin and sucked in a large hit. The smoke was clean: flavorless on the inhale, not even tasting like smoke. I exhaled and got a dry floral taste. Again, the smell was clean, not skunky or tasting like fuel or much of anything, to be honest.

I felt a brief skitter of THC into my head, but the high itself unrolled slowly, building at an almost imperceptible pace. After about fifteen minutes I was baked, but it was a clear and relaxing kind of baked. And, like the John Sinclair I'd smoked previously, it was unlike any marijuana I'd ever experienced. It wasn't about overpowering you, beating you down to the couch, or turning you into a drooling fool. There's a real clarity and focus that comes with smoking Sleestak resin, and yet it produces a mild psychedelic effect.

I found it oddly reassuring.

Aaron took a hit and relaxed. Our conversation digressed and we talked about what it was like to live in Holland. He was worried about where his daughter would end up going to school. If he stays in Amsterdam, he'll be staying for a long time. He's not going to yank her out of school once she starts.

"None of that military family bullshit for her."

He sounded like most new parents. I told him about a friend in London who'd already got her newborn daughter on preschool waiting lists.

"Yeah, I figure I've got two years and then I'll have to figure it out."

He was born and raised in Los Angeles and misses his friends and family. "I don't go out much here, bro. I don't go out with the Dutchies too often." .

"But Amsterdam is such a beautiful city."

Aaron nodded. "Don't get me wrong, bro. I love Amsterdam. I feel safe here. If I had this business in L.A., some fools would try and jack me. I'd have to keep one of those Beretta police-style shotguns behind the counter or under my bed, and I don't want to. In Amsterdam I can sleep at night."

A homesick smile crossed his face. "But I really miss Mexican food."

I knew what he meant. Why is it that in Europe I miss Mexican food, but in Mexico I never miss European food?

The next day Aaron was in the middle of an intense phone conversation. I didn't know what, exactly, he was talking about or who he was talking to, but he wasn't messing around. This is not to say he wasn't polite; he was polite, but direct. He doesn't have time for bullshit. The DNA Genetics office responds to a couple hundred emails a day from customers around the world. They bombard Don and Aaron with questions about soil, indoor versus outdoor growing, organic versus chemical fertilizer, lighting, drying, storing, and even the odd query seeking relationship advice. If they're not emailing, they're calling. While Aaron and I were talking, his cell phone never stopped buzzing. The seed business was booming.

I followed Aaron into the back office again. He was ready to give me a lesson on the basics of botany. Just as we sat down, an alarm on his iPhone erupted. He stood and announced that it was time for lunch and he really needed to eat. He was ravenous.

"When I'm in Amsterdam I never eat breakfast. I go back to L.A. and I'm eating first thing. I don't know why."

He shrugged and walked into the other room, a small pantry area, and returned with a freshly microwaved pizza. Aaron made sure Raf got a couple of slices then sat down and began the lesson.

"Having pure genetics for us is like having primary colors for an artist."

He took a bite of pizza and chewed thoughtfully before continuing.

"We have those primary colors in our vault, so when we want to play, you know, start making some new colors up, we can."

I like the metaphor of primary colors. It's clear, it's basic, and it helps me understand how they concoct the various hybrids that make up the seeds they sell. Red and yellow make orange, which has characteristics of both red and yellow but is actually its own thing. From there you can experiment further, making various shades, hues, and tones. But without the primary colors, you can't start. That explains why they expend so much time and energy collecting various strains from around the world, putting together an extensive genetic library of what's called "landrace" strains.

A simple way to look at landrace seeds is that they are wild seeds, from nature, from plants that have grown together in one location and have inbred through open pollination, creating a pure—or at least predominantly unadulterated—genetic line.

The tomatoes we call "heirloom tomatoes" today used to be the kind of vegetable you'd find everywhere. Different regions would have different varieties, but they were all open pollinated fruit that grew in gardens around the world. In modern agriculture in the industrialized world, most food crops are now grown in large, monocultural plots. The advent of hybridization and agribusiness caused tomatoes to become uniform, and the more unique, oddball types got pushed out of the supermarkets and onto the fringes, where they are primarily grown by home gardeners and specialty farmers. Some varieties have disappeared altogether.

The DNA Genetics vault is like the stoner version of the Global Crop Diversity Trust, an independent seed bank charged with protecting crop diversity and the historical record of plant genetics; or the Svalbard Global Seed Vault, which preserves an extensive collection of plant seeds from around the world in an underground cavern tunneled into the frozen rock of the Norwegian island of Spitsbergen in the arctic. The Svalbard facility is funded by a handful of countries—as various as Sweden, Australia, Ethiopia, and Brazil—and private trusts such as the Bill and Melinda Gates Foundation. Less than a thousand miles from the North Pole, the seed vault acts like Superman's Fortress of Solitude, preserving genetic diversity on the off chance the human race totally fucks things up.

I don't think the DNA Genetics vault is as elaborate, but it's just as important.

Aaron agreed. "It's very important, because at a certain point, nobody's going to have these genetics and everything is going to be crossbred and hybridized and all the same."

In other words, without underground botanists to preserve

and manage landrace genetics, cannabis could become a mono-culture crop. Can you imagine if all weed was the same?

Aaron popped another slice of pizza in his mouth.

"It's important to save the landrace. I'll give you a perfect example: Martian Mean Green. It won the Cannabis Cup and then the mother was lost to the police. But that doesn't mean the strain is lost forever because I have the original seeds. So when I'm ready to go looking in those original seeds and do a selection to find the Martian again, I can."

Aaron finished the pizza and leaned back on the couch. His expression changed and I couldn't tell if he was wistful or just digesting.

"Listen, bro, in this day and age nothing's that pure. Everything has got skunk in it and it's hard to find landrace genetics. Right now we only have two strains that are pure, X-18 Pakistani that's only been inbred to itself and the Mazar-i-Sharif Afghanistan seeds we sell."

"So what do you think? Is there a danger that all the genes will be contaminated? That cannabis could go the way of the industrial tomato?"

"Yeah. Lemme give you an example. It's definitely not hard to cross two plants together that aren't related and get a plant that looks and smells and tastes like a completely different plant."

"Like a third plant? Unrelated to the parents? Is that because the genetics have been muddied by years of breeding?"

Aaron laughed.

"It's weird, but it happens, because there's only so many terpenes and smells and flavors and tastes associated with cannabis. There's only so many combinations you can do. It's like tulips. There's a lot of different colors, but they're still tulips."

It may come as a shock, but Don and Aaron are not horti-culture or botany science graduates from the University of California, Davis or some other prestigious agriculture school. Like Floyd Zaiger—the California farmer who crossed plums and apricots to invent the pluot and aprium—they are self-taught amateur geneticists who follow their instincts. Because cannabis is dioecious—meaning it has male and female plants—the science they employ is strictly old-school. They don't splice genes under a microscope or genetically engineer plants by tweaking the DNA; they collect pollen from specially selected male plants and pollinate female plants by hand. They grow out those seeds, select, cross, and then grow and select again. It is a time-consuming, arduous process.

I'm curious. I get that it takes time, but what exactly happens? "How do you make a new varietal or strain?"

Aaron warmed to the subject. He started speaking faster, his voice rising, his eyes becoming more animated. This is a subject he loves. This is what he likes to talk about. "Let's look at Chocolope."

Chocolope is one of DNA's most popular strains and was named one of the top ten strains of the year for 2007 by *High Times* magazine.

"In 1989 to, like '93 or '94, a shipment of Chocolate Thai used to come over to L.A. every year. In those Thai sticks you used to find good seeds."

They hung on to the seeds for years, just saving them, keeping the genetics in the vault.

"When we came here, we decided it was time. We grew 'em out to see what was there."

"Is there anything particular you look for? Any early indication of dankness?"

"You look for one that flowers decent and doesn't look all scraggly. Most of the time you gotta look through a lot of shit before you find something special. Hopefully you find one; if not, you scrap the project."

"That sounds time-consuming."

Aaron nodded. "Yeah, bro, it can take months, years. But we grew 'em out and found a nice female and one male."

Chocolate Thai is a sativa with a flavor that accounts for the eponym and was very popular in the 1970s and '80s. Nowadays it's a strain that's considered old-school, heirloom, like a deep purple Cherokee tomato.

Don and Aaron used the male and female plants for a couple of years, making and selling pure Chocolate Thai seeds. The strain became a stalwart at various coffeeshops around town.

Aaron continued. "But honestly, Chocolate Thai wasn't all

that dank. It was just okay to us. So we decided it was time to introduce the Chocolate Thai female to another gene pool."

They decided to cross the Chocolate Thai with a varietal they'd developed called Cannalope Haze. They pollinated the female and got a crop of seeds.

"You're allowed to grow up to five plants here, so I cracked some seeds on my roof terrace. Of course I had more than five plants. Maybe I went a little overboard."

Aaron looked down and rubbed the back of his head. A mischievous smile formed on his face. "I'll be honest, bro. It was kind of like Mendocino on my rooftop."

At a certain point, Don told him he'd better get back to the legal limit. So Aaron made a selection of what he thought were the best-looking plants. "Selection is playing god, bro. I took twenty of them, brought 'em inside, and flowered them out."

And that's when he discovered that one of the Chocolate Thai x Cannalope Haze plants was special.

"This one really stood out. It had the most unique scent."

"You could tell by the smell of the plant?"

"Yeah, bro, you just get up in it and smell it. To me that's the big factor when I'm selecting. I'm not talking about smoking. It's just the way the plant smells. That tells me a lot right there, and this one was special."

He destroyed the other plants, keeping the special one.

"So what did you do next?"

Aaron laughed. "The next step is you gotta smoke 'em. Sometimes the plants that smell great don't taste great. But this one did."

They had a great smelling, great smoking plant, but the next step was to help the growers. The new plant was sativa dominant and needed a lot of space and a long growing time, so they backcrossed it with Cannalope Haze hoping to get a bushier, lower growing plant with quicker flowering times. They succeeded.

"And that's Chocolope. Those are the seeds we sell."

It's fun to get a botany lecture from a dude in a sports jersey as he scarfs down a pizza. Even though he peppered the conversation with horticultural words like "phenotype" and "genus," "terpene," and "cannabinoid," what he said made sense to me.

There's no danger of DNA Genetics producing a monoculture cannabis. Their interest is in diversity, in creating a variety of tastes and sensations, flavors and effects. The only way to ensure that is to continue developing very personal, handmade cannabis strains.

Because DNA Genetics made its name by winning the Cannabis Cup, Aaron and Don are serious about the competition. Their goal is to be in the mix, among the top competitors jockeying for the prize every year. I was curious what their strategy was. What they were entering in the Cup.

Aaron threw up his hands, helpless.

"We don't know. It's up in the air."

I was surprised. I had assumed that they were experimenting, planning ahead, trying to come up with something that would kill the competition.

Aaron shrugged. "That's one of those things. Every year, we wait until the very last day when we have to turn in our entries. We're not growing anything now; it's not time yet. We won't start until three or four months before the Cup."

I thought maybe he wasn't telling me the whole truth. He must have something up his sleeve, but Aaron shook his head.

"It's what happens to show up at the time. Last year we had some Headband, some #18, some Chocolope—really we don't know what it's gonna be."

"So your entry isn't based on the strain? It's based on the quality of the crop?"

"Yeah, it's whatever is the real quality herb at the time. I'm not going to enter some crap. I mean, last year we could've made a statement and said okay, we're not going to win anyways so let's just enter some dirty brown Jamaican weed and call it 'The Dirty Brown Jamaican' and see what happens. But we didn't do it."

Their track record at the Cannabis Cup—and other international competitions—has turned DNA Genetics into one of the few "top of the line" brands in the international seed market, a fact that Don and Aaron are clearly aware of. They're already looking to expand on their success, branching out into a clothing line and a collection of DNA-branded hand-blown glass bongs and water pipes produced by renowned glassblower Sheldon Black.

And then there's California. They're working with a lawyer to set up an operation that complies with medical marijuana laws already on the books. Aaron likes the idea of coming home, only this time he'd be legit. He stretched back on the couch.

"If we can get going in Cali and we can make a move without creating too much wind, I'd like to see us bounce into the other states that have medical marijuana laws."

He looked at me and smiled.

"And you know what, bro? We'd kill it."

Chapter Seven

The Underground

Franco walked into the coffeeshop dressed head to toe in yellow and black motorcycle racing leathers. The clothing was skin-tight, wrapping his skinny body in aerodynamic leather, the silhouette enhanced by bulging chest armor and pads on the knees and elbows. He looked built for speed, a cross between Italian motorsport champion Valentino Rossi and a gigantic hornet, and could've easily been mistaken for a hit man in a John Woo movie.

His face erupted into a toothy grin when he saw me, and he came over to give me a warm hug. "Have you been waiting long?"

I hadn't.

Franco had asked me to meet him for lunch at the Green House United Coffeeshop on Haarlemmerstraat. It's one of several coffeeshops owned by Green House and Green House Seeds, where Franco is a partner. I had been sitting under a vibrant gold-colored mural depicting what looked like the ancient Phoenician or Minoan alphabet and watching the ornamental carp swim back and forth in the massive fish tank installed under the glass floor.

It was early still, in between breakfast and lunch, and the only other customers in the coffeeshop were a young British man who was eating hard-boiled eggs with toast in between hits on a massive joint and a well-dressed man in his late forties who sat drinking coffee, working on his laptop, and inhaling from a vaporizer bag. Except for the cannabis being consumed, it could have been any cool café in any city in the world.

Franco was met by a friend of his, a handsome Italian hipster named Davide who looked a little bit like the gypsy reggae rocker Manu Chao. Davide had something he wanted Franco to see, so he joined us at a large table in the open mezzanine of the coffeeshop.

We sat down and Davide began pulling an assortment of strange items out of his backpack: a couple of leather pouches, a pack of cigarettes, a stick that looked like a firecracker punk, and a butane lighter. Then he ceremoniously removed an object wrapped in a crocheted cover. It was about the size and shape of a large dildo, and, for a second, I wondered if this was some new European fad that I hadn't heard about: collectible macramé dildo cozies. Davide untied the cover and pulled out a wooden pipe. It was not like a normal pipe with a bend in it; this was straight, like a miniature *vuvuzela,* the kind of horn you'd toot at the World Cup or on New Year's Eve.

Davide took a cigarette and began gently toasting the tobacco by running the lighter flame along its length.

Franco explained that the wide end of the horn had a special clay bowl in it that would produce very high heat for smoking *charas,* a kind of oily Indian hashish.

Right on cue, Davide opened one of the leather pouches and removed a chunk of deep brown putty. He crumbled some of the *charas* into one of the leather pouches and mixed the hash with the toasted tobacco.

Franco rubbed his hands together as Davide packed the mixture into the pipe.

"This is how we smoke in Italy. This is the traditional way."

Davide held the lighter to the wide end of the horn, cupped his hand around the narrow end and inhaled. The *charas* and tobacco mix glowed like a hot orange eye. Then Davide handed the pipe to Franco.

Franco took a big pull and offered the pipe to me. I declined. I don't like tobacco. Franco exhaled and smiled. "We used to smoke like this when we were kids."

He handed the pipe back to Davide, who took an experimental puff and realized that the bowl was cooked. He then used the stick that looked like a firework punk, ran it the length of the pipe, and poked the burned mixture out the end. He took a soft rag and cleaned the inside of the pipe as if it were the barrel of a gun. Davide grinned, like he'd just won a sharpshooting contest.

. . .

If you couldn't have guessed from his racing leathers and Ducati motorcycle, Franco is Italian. Like Don and Aaron, he's one of the younger generation of underground botanists making their mark on the cannabis world. But unlike other strain developers and seed growers I've met, Franco has the unique distinction of having served as a paratrooper in the Italian army. Jumping out of planes, racing motorcycles: He's not the kind of guy who likes to sit still—in fact, there's a lot to suggest that he has a need to be hurtling through space. I reminded myself not to take a ride on the back of his Ducati.

Italy is one of the more hostile environments for cannabis users—even though, as Franco says, "everyone smokes"—so after his compulsory military service was over, Franco moved to Amsterdam to attend college and earn a degree in hotel management. While he was in school he began growing pot for fun—experimenting with the plant for his personal use—and he quickly discovered he had more than a smoker's affinity for cannabis.

In 2001 he got a job as a manager at the Stichting Institute of Medical Marijuana, which, at that time, was the only licensed producer of medical grade cannabis in the world, growing exclusively for the Dutch government and pharmacies throughout Holland. With his experience as an organic grower and strain developer—and his taste for a fast-paced, adrenaline-fueled lifestyle—it was only natural that he joined Green House Seeds in 2004.

To borrow a sports cliché, Green House Seeds is the New York Yankees—or FC Barcelona—of the cannabis seed industry. They have won thirty-two Cannabis Cups, seventeen High Life Cups, and a bunch of other awards from competitions as far-flung as Spannabis in Barcelona and Cannabis Tipo Forte in Italy. More important, they are credited with creating some of the most popular varietals in the cannabis world: Lemon Skunk, Hawaiian Snow, Himalayan Gold, Super Silver Haze, White Widow, White Rhino, and the two-time Cannabis Cup champion Super Lemon Haze.

Green House Seeds was started in 1995 by an Australian

cannabis breeder named Shantibaba and a Welshman who went by the *nom de pot* Mr. Nice. Mr. Nice was, at one time, considered the biggest marijuana and hash smuggler in the world and eventually served a seven-year stint in a U.S. federal penitentiary. A third member was added to the company, a mysterious Australian—or mysterious Dutchman, depending who you ask—named Neville Schoenmaker. Neville is a bit of an enigma; even the spelling of his name varies from report to report. Sometimes it's Nevel, other times Nevil. It's hard to know if the typos are related to cannabis intake or some kind of intentional confusion that Neville himself perpetrated. However you spell his name, after the U.S. Drug Enforcement Agency put him on its "Ten Most Wanted" list, he disappeared in Indonesia or Australia, depending on which story you want to believe.

Neville, along with a Dutch South African named Arjan, was one of the owners of the Green House Coffeeshop at that time, but he is more famous for being the father of the Dutch seed industry and founder of the Seed Bank, one of the first cannabis seed clearinghouses.

Like all origination myths, there are several different versions of how the Dutch seed business began. All of the versions, interestingly, orbit around the introduction of a unique strain of cannabis called Haze into the world of underground botanists.

One version has a Californian named David Watson—who goes by the *nom d'weed* Sam the Skunkman—bringing Haze genetics to Amsterdam and sharing them with several breeders, including Neville. Another interpretation has Neville venturing to Santa Cruz, California, in the late '80s, where he encountered two local growers called the Haze Brothers. It was the brothers who introduced Neville to the powerful Mexican sativa.

Both Sam the Skunkman and Neville have contentiously stuck by their stories in different interviews, although I have to say, just for the sake of argument, that there's no reason they both can't be right. Sam the Skunkman could've given Neville seeds of Haze and Neville still could've gotten other Haze seeds from the Haze Brothers.

Naturally, Sam the Skunkman denies that Neville got the real Haze, alternately claiming that he gave Neville an inferior version of the plant or that the Haze Brothers gave Neville bad genetics. The truth is—dare I say it?—*hazy,* but the fact remains that Haze came to Holland and the cannabis world has never been the same.

Haze is an interesting sativa varietal. It has a unique spicy flavor and gives smokers that uplifting and clear, euphoric feeling that I refer to as dankness. It's a sensation that a poet friend of mine once described by declaring, "It makes me feel like I can understand algebra." It also seems to have a mildly psychedelic effect. It is, like most sativas, extremely difficult to grow and delivers a very small yield for the amount of space and effort required.

But what's interesting about Haze is its ability to play well with others. Crossing Haze with a more grower-friendly strain of indica not only produced plants that were reliable for growers, but also brought forth some of the best-tasting cannabis strains ever concocted. There is original Haze found in DNA Genetics' Cannalope Haze, Chocolope, Sour Cream, and C-13 strains, while Green House offers variations called Arjan's Haze, Alaskan Ice, Arjan's Strawberry Haze, El Niño, Jack Herer, and others, and Shantibaba's Mr. Nice Seedbank produces Haze-inflected strains like Afghan Haze, Mango Haze, and Neville's Haze. You can find Haze genetics in dozens of the most popular strains being produced in Amsterdam.

Green House Seeds crossed Haze with Skunk and Northern Lights, creating a strain known as Super Silver Haze, winner of the Cannabis Cup in 1998 and '99, and arguably the most popular strain ever produced.

But whatever story you want to believe, the introduction of Haze genetics kick-started the Dutch seed-breeding revolution and made Neville, among others, a millionaire.

In the early days, as the seed industry began to take off—partly due to the popularity of *High Times* magazine's efforts to promote superior quality marijuana—Neville, Shantibaba, and Mr. Nice began operating under the Green House banner and won every award at the 1998 Cannabis Cup. This victory was followed by money, prestige, creative differences, discord, arguments, ego trips, and the dismantling of the original team.

Shantibaba, in collaboration with Mr. Nice, formed Mr. Nice Seed Co., Neville vanished, and Arjan bought out his partners in Green House and brought his genius for marketing and brand identity to the business, turning Green House into the most successful cannabis company in the world.

To give some perspective, a midsized operation like DNA Genetics will sell approximately 500,000 seeds per year. Green House Seeds sells about 4 million, totaling more than 20 million euros a year in sales. But that's just a small part of their operation. Green House also owns four successful coffeeshops in Amsterdam, a profitable clothing line, and a real estate company that rents high-end VIP apartments to tourists. Green House has more than one hundred employees, and—I don't know why I find this so surprising—even has a human resources director.

It's no wonder that Arjan calls himself the "king of cannabis."

I met him briefly in Los Angeles at the THC Expose convention. We talked a little, and when I told him I was coming to Amsterdam for an extended period he said, "I don't live in Amsterdam anymore."

I knew that Arjan is originally from South Africa, so I asked him where he was living.

"I live in the bush."

I looked at him and raised an eyebrow. He shrugged.

"I like it."

I told Franco this story and he laughed. "If where you stay most of the time is where you live, then I don't think Arjan lives anywhere. He's always on the move."

Franco obviously has deep respect for his partner. "He is like a fountain of ideas," he said. "Everything comes from Arjan. The idea to put color on the seeds, the Strain Hunter movies, he's always thinking."

Green House puts a unique protective coating on their seeds with each strain color-coded so the grower can easily identify it. This way they can offer a variety of sativa-indica mix packs and even something called a "Rasta Mix."

Franco is a big advocate of the mixed-seed packets.

"I believe the one great thing about cannabis is the variety.

It's so boring to smoke the same thing all the time. Our brain receptors are made to enjoy the variety."

"But don't people have their favorites?"

He nodded. "For sure, but if you smoke different varieties you need less to achieve the same effect. You don't build up a tolerance."

This was news to me. I know that daily users can build up a tolerance to THC, just like someone who drinks every day builds up a tolerance to alcohol. But I'd never heard about mixing it up as a way to avoid that.

Franco waved his finger in the air to make his point. "That's why we encourage people to grow several strains instead of one. Variety is key. That's why we sell five strains in a mix pack."

These are feminized seeds, so that the amateur grower will be guaranteed to get female plants—the ones that produce the THC-laden blooms—and because the seeds are color coded, they'll know which plant they've planted. I've heard other seed companies mock the colored seed concept, calling them "Fruit Loops" and saying "Trix are for kids," but it really is an ingenious way to take the guesswork out of the process for the amateur grower. And home growers are the customers who account for a large portion of Green House's sales.

Franco knows I have a fondness for sativas so he pulled out a strain he called Special Sweet Skunk.

"People who need to perform in their lives, for sure they prefer sativas. But when I like to relax, at the end of the day, then usually I like an indica like Cheese."

"I like a glass of wine at the end of the day."

Franco grinned. "That's good, too."

He slid a small plastic box toward me. Inside were a dozen chunky nuggets of Special Sweet Skunk. I inhaled and got a very clean and fruity, almost Hawaiian Punch flavored, scent off the buds. They smelled pungent and lively, just like dank cannabis is supposed to smell.

While Franco rolled a fat spliff, I watched Davide put his *charas* smoking apparatus away. He handled each object reverently, moving slowly and precisely, like a Santeria priest storing his fetishes.

Franco lit the spliff, took a drag, and handed it to me. I took a big hit and felt my lungs explode. They didn't explode in a pleasant way; they exploded in an explosive way. I coughed. And coughed. And then I really start coughing.

The fruity sweet smoke I was expecting had been replaced by the burnt-leather taste of tobacco. The aftertaste hit me like a bad memory, specifically, being in college and making out with this punk rock girl who smoked a lot of cigarettes and wore cherry-flavored lipstick. It was like licking out an ashtray and discovering an old Lifesaver stuck to the bottom.

In between convulsions, I looked at Franco and raised an eyebrow. "Tobacco? What the fuck, Franco? I'm from California."

Franco burst out laughing. "Oh, man. I'm sorry. I'm really sorry."

Personally, I don't know why Europeans like to mix tobacco with their cannabis. For me it defeats the whole taste and smell pleasures of a nice weed or hash. It's like pouring a Miller Lite into your pinot noir.

"I'll roll you a pure one."

Franco laughed some more, then looked at Davide. They both took a deep breath, and then erupted in laughter again. I didn't see what was so funny, but the Italians found it hilarious. Franco doubled over, pounding the table with his palm.

"So sorry, man."

"It's just that we don't smoke tobacco in California." I really hoped that didn't come out as whiny as it sounded.

Franco took a hit off the tobacco and Sweet Skunk spliff to stabilize himself. He looked over at Davide.

"That's right. When I was there, in L.A., I wanted cigarettes and I had to walk five blocks to find a place. Meanwhile I'm passing two or three marijuana dispensaries."

They continued chuckling while I swished my mouth out with some ice water. If I could've brushed my teeth at the table, I wouldn't have hesitated.

I wanted to talk to Franco about Super Lemon Haze. It was the reigning Cannabis Cup champion, having won two years in a row, and one of the most popular strains in the world. I had tried some at the previous year's Cannabis Cup and I liked it a lot. It has a mild citrus flavor and a clear sativa high. It's easy to see why it's such crowd pleaser. I was curious how they developed it.

"Well, it's a very old-school classic taste with a new twist. The father is based on Super Silver Haze and the mother is Lemon Skunk."

Franco leaned forward. He wanted to make sure I heard this right. "Lemon Skunk is a very special plant. Combining the two has been a natural success story. And actually the breeding was very easy. We just made an F1 hybrid from the parents, planted a bunch of them, and made a selection. Select, select, select. That's the key to finding something amazing. It was amazing to me that you can take out a champion from just an F1 hybrid."

"F1 hybrid" is a term used in genetics that stands for the first generation of offspring from two distinct species. For example, in the case of Chocolope, to get the right characteristics, Don and Aaron had to backcross their F1 hybrid with the mother plant to create an F2. And while it's not unusual to have a good F1 hybrid if both the parents have good genetics, it is rare to have a world-class F1 hybrid just pop up.

"What gave you the idea of combining the two plants in the first place?"

Franco began rolling me a pure spliff of Special Sweet Skunk while he talked. I tried to listen and, at the same time, see how he did it. He's obviously had a lot of practice because he did it without looking. The former paratrooper could probably roll a joint while jumping out of an airplane and still carry on a conversation.

"Creating strains is something that requires three things: landraces to start with, logistics—very big logistics are required—and it requires good knowledge of the customer base . . . what people want."

"Like any other business."

Spliff rolled, he licked the paper and sealed it, then continued.

"Green House the company is in a very special position in this respect. We are one of the very few seed companies that have access to a large customer base through our coffeeshops. We have thousands of people coming over every day from literally all over the world. It's very easy for us to get feedback and to see what people smoke here, what they like, and why they like it."

"So you, what? You run coffeeshop focus groups?"

Franco nodded.

"We knew that citrus was a new trend. We knew it three years ago when it started. Every time we add on our menu something slightly citrusy, all of a sudden people were raving about it. It was a little niche flavor in cannabis. And we felt it was ready to go. So we took a very old flavor, a sativa flavor—a flavor that people were almost sick of—and twist it with something that makes it new again."

He handed me the joint. I lit up and this time I didn't cough. The smoke tasted as good as the bud smelled. Absolutely delicious.

I assumed that Super Lemon Haze would be the natural choice to enter in that year's competition; after all, no strain has ever won three years in a row. But Franco wasn't so sure.

"What we do is, every year for the Cannabis Cup, we wait for the last minute to decide what we're gonna enter. Because we like to cure our crops and it's hard to predict how the crop will cure. So we always have two, three, four crops that we know are very high quality and depending on how they cure, we can make the final decision."

Just like Don and Aaron of DNA, the final decision is more plant based than anything else.

Franco nodded. "Of course, there are several strains that we want to promote, and we are going to make sure that those strains are going to be in the final mix. But only after curing can you be sure you've got a winner."

"How can you tell what will be a winner?"

Franco grinned.

"It's what will make people flip out when they see it."

Personally, I was flipping out for the Special Sweet Skunk. It was one of the best strains I'd ever tasted.

I looked at Franco as I exhaled. "What does 'dank' mean to you?"

He took a sip of an espresso and considered it. "Dank is a very American concept so to be defined by a European is already a tough job."

"Isn't there some European version?"

He thought about this for a moment. "The European version of dank? I guess it would be 'the underground.'"

I wasn't sure I followed, but Franco started getting into it, talking rapidly with lots of Italian hand gestures. "Listen, dank is a way of life, it's a way of living, being connected with the plant—a plant that has been disconnected from the people for a long time because of its illegality, because of its status. A lot of people want to reconnect with this plant because they feel the plant has been taken away from them. And so people unite in one feeling, one community. Some of them do it recreationally; some need it for medicinal reasons. Some for religious purposes because they feel they can talk to God with the plant. There are many different reasons, but the bottom line is that humans have been in relations with this plant for thousands and thousands of years and you cannot just break the relationship by declaring the plant illegal. It doesn't work like that."

So "dank" means you're a member of a community? "I thought 'dank' had more to do with the quality of the plant," I said.

Franco took a hit off the tobacco and Special Sweet Skunk joint and wagged a finger at me as he exhaled. "'Dank' goes beyond nature. This plant is proven to unite civilizations. It's one of the few common natural elements of mankind. It's a common link throughout history, throughout geography. You can connect humanity with this plant. Asia, Africa, down to the Americas. Everyone has been using this plant for one reason or another throughout our history. Name another plant like that."

I wanted to say wheat or rice, but then I wasn't sure. It's true that evidence of cannabis and hemp use goes as far back as recorded history, and it's been used in almost every culture for rope, fabric, medicine, and spiritual practice. And besides, it was hard to argue with an impassioned former paratrooper dressed in leather body armor.

"This plant has been taken away and now people want it

back. So being dank, living dank, this is a way that allows you to get back to the plant."

Franco stopped and burst out laughing. "Or maybe it allows the plant to get back to you."

"Living dank? What does that mean?"

"In Europe we have a big underground movement that is very dank, for sure, people who have very little in common apart from this plant. It goes through all of Europe, it's rich people, poor people, cultivated people, sporty people, couch lock people. That's the funny thing about it. It's a very powerful connector."

"Like a true underground."

Franco nodded. "Because the plant has been illegal for so long, everyone who has ever tried it knows or has had the feeling of doing something wrong. But when they try it, they don't feel it's so wrong. It's natural. It's a plant. What can be so bad using a plant? This discrepancy you get from knowing you're illegal but feeling good about what you're doing, that you're not doing anything wrong—this feeling creates a bond between people."

I think I know what Franco's talking about. That bond is the glue that holds the counterculture together. It's easy to take Franco's description and stretch it into the realm of lifestyle and politics. The discrepancy between what a government decrees as bad and what a person understands on a deep, intuitive level as good, is not restricted to our current marijuana laws. Modern society oozes with this kind of hypocrisy; it's reflected in the way we talk about sex, the way we educate our children at school, how our politics are disconnected from the populace, and the list goes on. There is a growing gap between those with the ability to think critically and follow what their intuition tells them is the right thing to do and those people whose morality has been perverted by greed and short-term corporate gain.

Franco knocked back his espresso and stood.

"Time to go back to work."

Chapter Eight

We Circumnavigated the Globe and All We Brought Back Was a Grilled Cheese Sandwich

One of the things people don't come to Holland for is the food. No matter how fresh baked and darkly grained the bread is, no matter how pure and wholesome the cheese or how savory the meat, there are only so many variations on the grilled cheese sandwich a person can eat.

They're called "toasties" or *broodjes,* and they are ubiquitous. Even the cookie Holland is justifiably famous for, the *stroop-wafel,* is just two thin wafers on either side of a blob of caramel that's been flattened on a grill. It's just a small, sweet toastie.

As Jon Foster of Grey Area eloquently put it, "The food here is not dank."

I don't get it. Honestly. At the height of their empire, in the sixteenth and seventeenth centuries, Dutch ships ruled the seas. They sailed to Ceylon, Indonesia, the Spice Islands, South America, North America, Africa, and all points in between. There were more than twice the number of ships sailing under the Dutch flag than the English and French combined. They brought back coffee, tea, silk, and spices. They formed the Vereenigde Oost-Indische Compagnie, or Dutch East India Company—one of the first, and certainly not the last, large corporations to wage war, fix prices, and otherwise exploit and plunder at will—and

Amsterdam was home to the first commodities market in the world. Even New York City was originally called New Amsterdam before the British snatched it from the Dutch.

The Dutch ruled the world for a couple of hundred years and the best they could do is a toastie? Look at what other European countries were able to accomplish. Marco Polo goes to China, returns to Italy with some noodles, and the world gets pasta. The Spanish conquistador Hernán Cortés goes to Mexico and brings back the tomato. Gazpacho, marinara, and the Bloody Mary soon follow. Even the British developed a taste for a nice vindaloo.

Aside from the grilled cheese sandwiches and the occasional nibbling of raw, pickled herring on the streets, the Dutch tend to eat things like *bitterballen,* a small croquettish ball of fried meat; and *hotchpotch,* which is essentially just like it sounds, a pile of boiled meat and sausage on top of mashed potatoes and carrots. Hearty fare for active people. And, for what it is, it's not terrible. *But would it kill them to use a little salt and pepper?*

Amsterdam is a multicultural city, and it has some good ethnic restaurants. I ate reasonably tasty Indian and Chinese food and had some interesting Surinamese dishes, including peanut soup, which tasted like molten peanut butter with some chili oil mixed in. There are, as you'd expect, a few really good restaurants, but these tend to serve Frenchified nouvelle cuisine and are too expensive for a writer on a budget.

But I did find food that was delicious and cheap. It was at a simple burger stand called Burgermeester.

I know you might think, "Well, of course, the American wants his burger," but that's not it at all. While I'm not a vegetarian, I'm not a big carnivore either. I like to mix it up. I didn't eat the "Meester Biefburger" or the "Biefburger Royaal"—which is made majestic with the addition of truffles and pancetta. I will admit I found the duck, lamb, and salmon burgers all pretty good, but the one dish that I found absolutely delicious was the "Manchegoburger": essentially a veggie burger with Spanish cheese, hazelnuts, and a quince compote on it.

I don't think the lack of flavorful food in Holland is some kind of imperialist oversight or neglect on the part of the Dutch sailors; I think the Nederlanders really like their *broodjes.* *Broodjes* pair well with beer, and at brewing beer, the Dutch have staked a serious claim as masters of the art.

Heineken is Europe's largest brewer and one of the most popular beers in the world. Everyone recognizes the green glass bottle with the little red star on the label. It's like Pepsi or Coca-Cola—you see it everywhere. According to the Heineken Brewing Company's 2009 Annual Report, their worldwide consolidated beer volume was more than 125 million hectoliters of beer a year. I don't really know what that means, because I suck at ciphering and other mathematical endeavors, so I asked world-renowned mathematician Dr. Steven Wegmann to break it down for me. "The idea is very simple. For example, how many liters are in 125 million hectoliters? The units multiply in the following way: hectoliters times liters per hectoliter equals liters, so 125 million hectoliters multiplied by 100 liters per hectoliter equals 12,500 million liters. Then the next step is to convert the liters into pints using the same formalism: 12,500 million liters multiplied by 2.11 pints per liter equals 26,375 million pints. So how many pints of beer did Heineken produce? 125 million hectoliters times 100 liters per hectoliter times 2.11 pints per liter equals 26,375 million pints, or over 26 billion pints."

Sounds easy when he says it. I think, for the layman, it's mathematically correct to say that Heineken makes a lot of fucking beer.

The funny thing is, I never really liked Heineken. It tasted kind of skunky to me, and not in the good way that some cannabis tastes skunky. I didn't hate it, but it was never my first choice. But then I came to Amsterdam and found myself in a cozy little pub that only served Heineken.

The taste of fresh Heineken, unpasteurized, unfiltered, unwhatever-they-do-to-it-when-they-fuck-it-up-and-ship-it-to-the-United-States, was a revelation. It's not skunky; it's yeasty and lively. In other words, it has flavor. It is especially refreshing after a good smoke in a coffeeshop. Could a cold Heineken be an element of dankness?

Maybe that's what the Dutch explorers discovered. Because when you're buzzed, a beer and grilled cheese sandwich really hits the spot.

Chapter Nine

A Party for
the People

I didn't really expect the cannabis seed business to be so lucrative. Of course, now that I think about it, try telling that to Burpee or Monsanto. The trade in cannabis seeds isn't quite as big as those kings of agribusiness, yet it is still a multimillion-dollar-a-year endeavor. Make no mistake, Green House Seeds, DNA Genetics, and the other big seed companies—Sensi Seeds, T.H.Seeds, Barney's Farm, Kiwiseeds, and Dutch-Passion Seeds, to name a few—are all jostling for a share of a very robust market. While these companies develop their own strains, there are dozens of secondary seed brokers, like the Attitude, a "Cannabis Seed Superstore" operating out of England, the Vancouver Seed Bank in Canada, and Sweet Seeds in Spain, that act as retail outlets for seed companies.

It's unclear just how big the market is because the majority of transactions are quasi-legal and exact numbers are difficult to pin down. The 2006 U.N. World Drug Report suggested that an estimated 164 million people worldwide use cannabis regularly and that the global market for cannabis and cannabis-based products ranged from $10 billion to $60 billion annually. I think that's actually a conservative estimate. In 2011, a study done by an independent financial and information firm called See Change Strategy estimated the "national market for medical marijuana was worth $1.7 billion in 2011 and could reach $8.9 billion in five years." Another study, this one by Jon Gettman, who holds a Ph.D. in public policy and is a former national director of the National Organization for the Reform of Marijuana Laws (NORML), includes illegal sales and estimates that the U.S. market alone is worth $113 billion a year.

How much of that money trickles back to the seed companies? $100 million? $200 million? More? The seed companies themselves aren't saying.

I stopped into DNA Genetics and found Aaron in good spirits.

"We've had a good month, bro. We fuckin' crushed it."

They had received new seeds from six of their most popular strains, seeds that had been "out of stock" for almost a year. Now that they had fresh stock, customers weren't waiting to see how long the supplies would last.

Aaron beamed. "We probably outsold Arjan."

He pulled a bobblehead doll of Arjan, the "king of cannabis," out from behind his counter. I had to laugh. I always thought bobbleheaddom was reserved for athletes and politicians.

"They made a bobblehead to look like Arjan?"

"Yeah, bro. They got everything over there. Arjan's a celebrity."

"Where's your bobblehead? You guys are celebrities, too."

Aaron shook his head.

"We don't look at it like that. My friends are like 'Dude, you're a celebrity, homie.' But I don't feel like that. I'm just Aaron of DNA." He shrugged. "We do what we love."

Aaron whacked the doll and the head vibrated. He laughed.

"I tried to drown Arjan. We put it in the aquarium and it started to swell and get all deformed, but the paint that he's made with is toxic or something and we lost a few fish."

Aaron shot a guilty glance at the fish tank.

"Then Don tried to blow him up with fireworks, but, look." He shook the bobblehead again. The king of cannabis's head spun around and waggled violently before settling into an agreeable nod. "He's fucking indestructible."

"You really should make Don and Aaron bobbleheads."

Aaron looked at me and made a face like he'd just tasted something bad. "Then what? Should we call ourselves the kings of cannabis? The princes of pot? Should I name a strain Fuckin' Aaron's #1? I don't want to do that."

Arjan has a tendency to name strains after himself: Arjan's #1 or Arjan's Ultimate Haze.

Aaron pointed to the bobblehead. "Look, I like Arjan. Arjan is a good guy. The world might think he's some egotistical guy, but they don't know him. He's a good guy. He's the king of marketing. I just look at him and watch and learn."

As Arjan's bobblehead bobbled I asked Aaron why they ran out of seeds. He looked at me like I was a moron.

"The plants, bro. We're one of these companies that are sold out of our most popular strains. You look at some of these other companies and they're never sold out. I don't know how the fuck they don't run out of seeds. Maybe they're filling in the packages with other seeds."

Since most growers won't know what they've got until it's grown, it's easy for some dishonest labeling to occur.

"You got Spanish companies that all sell the same fuckin' shit now. They're all selling the same seeds made by three different companies."

Aaron shook his head, disgusted.

"I don't want to be that guy. If we're sold out of a strain . . . " His voice trailed off as he considered what he was saying. "We haven't had Martian Mean Green in over two years. I haven't had Cannalope Haze in a year."

These are huge sellers when available.

"I'd rather be sold out than sell bullshit. But a lot of these companies . . . they do that. It's scandalous."

He whacked the Arjan doll again. The king of cannabis's head rocked back and forth, agreeing with Aaron.

"It's a seedy business, bro, no pun intended."

Almost all the big seed companies, and many smaller ones, sell their seeds by advertising in magazines like *Skunk, High Times, Weed World,* and *Treating Yourself*—a magazine that focuses on the medical side of cannabis use. Because it's illegal to send marijuana seeds to the United States, all of them have disclaimers stating that they don't take orders from the States and that they don't sell seeds to customers in America.

And yet the seeds somehow find their way to the American

market. L.A. Confidential, Trainwreck, Chocolope, and other strains from Dutch seed companies are sold at several dispensaries in Los Angeles. Those seeds are either purchased in England or in Holland and then carried into the country, or the sales are handled by third-party websites and smaller seed dealers who don't mind taking a calculated risk. As David Bienenstock, the senior editor at *High Times,* told me, "It's a business totally based on trust and reputation. You stick some cash in an envelope and send it off to Europe hoping you'll get something in return."

But for quality-control freaks Don and Aaron, turning your seeds over to another retailer is not always a great idea.

"The danger is you can send your seeds to some guy who has a little website and he doesn't send you a fucking dime. Then he says, 'Oh, some shit happened,' and then you're fucked because it's not a legal transaction."

Aaron shook his head. "Last year at the Cannabis Cup one of these small seed companies tried to jack us and I caught that fool. Dragged his ass outside and said, 'Give me your shoes.'"

Taking someone's shoes and then throwing them over an electrical line is a uniquely Los Angeles way of settling a score. I decided not to remind Aaron that there aren't any electrical lines crisscrossing the streets in Amsterdam.

"He didn't pay us, then he returned seeds that weren't ours and said they were. And then he insulted my wife." That was, obviously, a mistake. Aaron shook his head in dismay and I knew what he was thinking: It's hard to believe people can be so dumb.

He heaved a sigh. "I felt bad about it, in some sense, but if I didn't do it, he'd have kept doing it to other people. He needed to learn his fucking lesson."

Aaron stroked his goatee, reflecting on the incident.

"It was at the point where this motherfucker's shoes were off and I was gonna take 'em and Don said . . ."

Aaron paused, letting the drama build, and then did a perfect imitation of his partner. "Don said, 'Let's give him back his shoes.'"

Don's the softy of the two.

"We used to be on all these little websites but you know what? I'm done. I have a legit business now."

In 1987, the then editor of *High Times* magazine, Steven Hager, came up with the idea of a harvest festival to celebrate the new cannabis varietals that were being developed. It's not like there aren't other annual competitions for various food and drink around the world. In addition to the Concours Mondial de Bruxelles, the mega wine event with tastes of more than six thousand different wines, there's the World Beer Cup, known as the "Olympics of Beer Competition"; the Superior Taste Award, which judges food and drink in 270 different categories; and my personal favorite, the World Cheese Award, "the biggest and most cosmopolitan cheese festival ever staged." There are hundreds of local and regional competitions around the world for everything from hot sauce to apple pie. And those are just the contests that are judged using taste and smell; there are a multitude of other types of contests seeking excellence in human endeavors from precision origami to best pointless steampunk invention. There must be some kind of genetic predisposition that humans share that makes us want to see who makes the best whatever. But Hager knew that there wasn't an international competition for cannabis and that Amsterdam, with its tolerance and coffeeshops, was the natural place to hold the event.

It wasn't always the weeklong cannabis carnival that it's become. In the early days there were only a few seed companies that entered—breeders like Neville and Sam the Skunkman— and the celebrity judges were *High Times* staffers and the artists who drew the *Fabulous Furry Freak Brothers* comics.

But Hager had obviously struck a chord, and the Cannabis Cup took off. A few years later the "Coffee Shop Crawl"— basically the same idea as a pub crawl where you go from coffeeshop to coffeeshop across the city sampling their entries— was added as more and more strain breeders and coffeeshops joined in.

I don't think the amazing growth in the seed business could've occurred without *High Times* magazine and the

Cannabis Cup. Nowadays, of course, wannabe growers can get all kinds of information about seeds, genetics, and quality of the plants on the Internet, but in the halcyon days of the industry, winning the Cannabis Cup was the only surefire way of getting your strain and your company—your brand—out into the world. And if you were good enough or lucky enough to win? Your reputation—and the money that followed—was assured.

With so much money and prestige at stake, how intense is the competition?

Each year the big seed companies spend tens of thousands of dollars advertising in *High Times* and producing gift bags and parties and events for the competition attendees. Green House and Barney's—coffeeshops that are also seed companies and so promote their own in-house brands—offer free grinders, T-shirts, tote bags, and, of course, free samples of their strains. DNA Genetics doesn't have its own coffeeshop, but hosts an annual "Hot Boxxx" party that, in 2009, featured reggae superstar Barrington Levy. All of this is done to get the judges' attention, show them a good time, and promote the brands. Anyone who comes to the Cup can purchase a "judge's pass" that allows them to vote on the strains and hashish that the various coffeeshops have entered.

Coffeeshops typically enter a strain of marijuana, a type of imported hash, and a Nederhash, hashish that's made in Holland. If a coffeeshop doesn't have an in-house seed company, they usually align themselves with a seed company that doesn't have a coffeeshop. This is, basically, how everyone except Barney's and Green House operates.

In 2009, at the Twenty-second Annual Cannabis Cup competition in the Import Hash Cup category, first place went to Green House for Rif Cream, second place to Barney's for Triple Zero, and a coffeeshop called Amnesia showed for third with Azila. In the Dutch Hash Cup the results were flipped: Barney's placed first with Royal Jelly, Green House second with Green House Ice, and Jon Foster's Grey Area came in third with Grey Area Crystal.

The main event, the overall Cannabis Cup competition for best marijuana strain, revealed a similar domination by the big coffeeshop–seed company conglomerates. Green House took first place with Super Lemon Haze, Barney's was awarded runner-up status with Vanilla Kush, and a smaller coffeeshop called Green Place took the bronze with a DNA Genetics strain called Headband Kush.

Twenty-nine coffeeshops had entries in that competition. What that means is that a truly dedicated judge, someone who wanted to sample all of the entries before making a decision, would have to smoke eighty-seven different types of hash and marijuana in the five days of the competition. That's ingesting seventeen different types of cannabis a day. I'm not saying it's impossible—just like it's not impossible to climb the highest mountain in the Himalayas or run the one-hundred-meter dash in 9.58 seconds. But it's not for the faint of heart. It's a good thing cannabis is a nontoxic plant; it can't kill you no matter how much you smoke.

While the popular vote decides which coffeeshop and strain wins the overall Cannabis Cup and Hash Cups, there is a competition among the various seed companies for the best indica and sativa strains. These categories are called the Sativa Cup and Indica Cup and are decided by a panel of experts called the Temple Dragons—which includes most of the *High Times* staff—as well as some select "celebrity stoners."

Aaron laughed when I ask about the Temple Dragons. "I actually have a Temple Dragon T-shirt and, you know, when I put that on I'm in an elite, secret society."

I told him it sounded like something out of a Bruce Lee movie. He shook his head.

"It's not like that, bro."

I think it's safe to say that most of the judges—I'm talking about the average folks who attend the competition and make up the popular vote—aren't able to sample all of the strains and hash in the competition. So is this where the marketing of the big coffeeshops starts to affect the vote? I asked Aaron what he thought.

"I don't know, bro. Sometimes I don't understand how they win. I'll use last year as an example. Barney's won [second place] with Vanilla Kush. Now if anyone who was from L.A.

was here, they would know that Vanilla Kush had nothing to do with Kush. It didn't smell like any kinda Kush. I mean, I don't see how it was a winner. Personally, it wasn't dank. Forget me being a breeder, just as a connoisseur, I would never have selected the plant. You know?"

He held up his hands in exasperation.

"There was nothing there. But you have three hundred people that you give free herb to and you give 'em free goodie bags and you know, maybe it's not always about the herb."

But then he changed his mind. "But then you know new people come in and win the Cannabis Cup too, it's cool."

I discussed this with cannabis activist and connoisseur Debby Goldsberry, a celebrity judge at the 2009 Cannabis Cup. According to her, the Temple Dragons are scrupulous about testing the entries. Strains are given code names and no one knows which coffeeshop or seed company has entered which strain. It truly is a blind tasting, just like the best wine competitions in the world. The public judges, however, don't taste blindly; they go from coffeeshop to coffeeshop to sample the various entries. Like me, Debby had wondered if the marketing efforts of the big coffeeshops and seed companies skewed the results, but she said that the "blind test came very close to the public test."

I told Aaron this story.

"It's funny that you say that because they did the blind tasting and everything and I hear this from *High Times* staff that when they left the hotel after the tasting we had supposedly won the Cannabis Cup."

Aaron leaned forward, rubbing his hands together. I could tell his sense of outrage was warming up. "And then when they get to the awards show, we don't win the Cannabis Cup."

He shook his head in frustration.

"So, I'm not too sure about it. I don't know. When you win, you win. Now three years ago when Chocolope lost by one vote . . . I think that was total bullshit and I'll say that straight up. It lost by one vote to Barney's?" He gave me a look of complete and utter exasperation.

"Come on now. And they have on film them taking away

two votes? Two Japanese guys were going to vote for Chocolope and then Barney's gave them a bunch of herb, so they voted for them. They sold their votes. Now, was it real? I don't know. But there is no fucking way you lose by one vote. Especially when the whole town came up to us and told us how good Chocolope was. Even people who got free judge's passes from Barney's—'cause, hey, that's how it works—even those people were voting for Chocolope."

He put his hands over his face and took a deep breath. I could see him telling himself to calm down. He looked up at me.

"Look, we believe that at the end of the day it's about the quality of herb. That's why we've lasted so long here in Amsterdam. That, and we haven't fucked anyone over."

It's not just seed companies that battle for the glory. Coffeeshops can make or break their reputation in the Cup.

Jon Foster has a unique perspective. Grey Area is almost always in the mix and has won the Cup several times, and yet it doesn't have the marketing budget of a successful seed company behind them.

"The key with the Cup is the popular vote. The big coffeeshops, I call them corporate sponsors, have seed companies connected to them and so they enter their own strains and that's a big commercial side of it for them. But we don't have that and for the most part we work with DNA and enter one of their strains. The thing with the Cup is, of course, that it's commercialized and promotion has a large part to play in the winning."

He adjusted his glasses before continuing.

"And you know because we're so small, we just can't get the number of people through our shop—you know, give them the whole experience like those other places do. I definitely see that as a factor that could hold us back from winning, even if we have the best strain."

Green House, the perennial champion, is one of the big sponsors of the Cannabis Cup, and they don't hold back. Last year I found Franco working every day at the Cup Expo—a trade show for seed companies and other cannabis-related entrepreneurs—standing in the massive Green House booth, filling up a fifteen-foot-long plastic bag with freshly vaporized Super Lemon Haze and spraying free samples out to anyone who put their nose up to it. But Green House wasn't the only company that did that. Almost all the seed companies with a strain in the competition were giving away free samples.

For Jon Foster, the Cup is less about competition and more about celebration.

"Some of the Dutch shops get really serious and they really try to win so they can say they have the best weed. But for me there is no best weed. There is x number of the best of the best— sometimes from small shops who maybe didn't get a mention or didn't get the traffic because they're too far out."

I'm curious about some of the smaller seed companies and coffeeshops who enter. What's the level of diversity among the various strains that get entered in the Cannabis Cup? Aaron offered some insight.

"We're friends with some of the *High Times* staff and after the competition's over, they'll come by the shop and show us all the entries. We get to sift through every single one."

Is there a trend that stands out?

Aaron shook his head.

"You can see how much crap is really entered. There's not so much weed where I'd go, 'That's good' or 'That's interesting' or 'I want to smoke a joint of that.' I don't know if it's because I only smoke resin now or just because a lot of the entries just aren't so special. Because when they're really special, you don't see them because they got all smoked up."

Most European countries allow their citizens to purchase and import cannabis seeds. For example, in Germany you're allowed to keep them as "souvenirs," but it's against the law to germinate them.

Dutch law doesn't make it easy for legitimate businesses like DNA Genetics or Green House Seeds. They used to be allowed to produce their own seeds in greenhouses and grow rooms in Holland, but the laws have changed. Now, while it's perfectly legal to sell and export seeds, and the Dutch government is happy to take the tax revenues that the sales generate, it's no longer legal to produce cannabis seeds in Holland.

Aaron explained how DNA produces their seeds.

"I don't make all our seeds. I contract people to make them. I got friends in Switzerland and Spain who do it for us. I can't legally do it myself, so I gotta have people do it for me. I gotta have a team."

This change in the law is not without risk for companies who don't have patent or copyright protection on the strains they develop. And even with patent protections it's hard to imagine a major pharmaceutical company like Merck trusting another company to produce their formulas.

"I give them the genetics and then I pray that they don't sell our seeds to other fucking companies like Barney's."

Aaron's not just paranoid. This is exactly what happened to DNA a few years ago when one of their growers sold seeds of L.A. Confidential to their competitors. According to Aaron, that version of L.A. Confidential went on to win a Cannabis Cup for Kiwiseeds.

"It was scandalous. They called it Mt. Cook. But that ain't no Mt. Cook. It's Mt. Crook. They should be fucking disqualified for entering it." He sighed. "But it happens all the time."

He was clearly angry but didn't blame his competitors as much as he blamed the grower who sold him out. "In California, you don't do that kind of shit to people . . ."

Aaron caught himself.

"A lot of bad things went through my head, but out here, it's changed my mentality, you know, and I was just like, let's just let karma take care of this motherfucker."

To be honest, I'm surprised that there isn't more corruption in the seed business. It's a relatively small world and there is so much money at stake. Aaron doesn't see it that way.

"I don't think we should be against each other. We're all in the same fucking boat."

There have, over the years, been accusations of vote fixing and influence peddling among some of the bigger coffeeshop–seed company conglomerates. In 1999, at the Twelfth Annual Cannabis Cup, Arjan and Green House were accused—along with two other coffeeshops, Rokerij and Het Kruydenhuys—of voting irregularities and were stripped of their awards in the Hash category. Steven Hager defended the competitors, claiming that because there were no written rules about how ballots should be filled out, it wasn't cheating so much as "confusion about what was proper and improper."

From my experience at the Cup I can honestly say that I didn't see any vote rigging. What I did see was similar to the campaigns run by Hollywood studios during Academy Awards season in Los Angeles. There were advertisements everywhere that screamed "for your consideration," private parties and VIP events where guests were handed liberally stuffed goodie bags, and, of course, lots of free samples. After all, an Oscar or a Cannabis Cup is worth its weight in gold.

I asked Jon Foster about the accusations that have dogged the Cup.

"A lot of people mention in the past that the Cup is bought, or the Cup is a hoax, but I think to put it in those words is too simplistic for what goes on. For me, the critical side is *High Times* provides the venue and puts in some impetus. But without what I call the corporate sponsors it would be a lesser event. They make it a really good time for the judges, and those are the people that it's about. For us, winning or losing is not really important."

"But it's better to win, right?"

Jon shook his head. "What's important is that the event continues and has a good name to it. It should just be a party for the people."

Chapter Ten

Strain Hunters

I liked Super Lemon Haze. I mean, seriously, what's not to like? It had been the people's choice, the overall favorite, the big winner the past two years in a row. I liked it, I really did. But—and this will ring a bell with many of the women I've dated in the past—I liked it; I didn't love it. For me it didn't have that other dimension, that hard-to-define quality that the John Sinclair, Sleestak, Zeta—or even the Special Sweet Skunk—strains had. But then when everybody was raving about California chardonnay I was more interested in drinking New Zealand sauvignon blanc and Italian arneis, so it could be that I'm suffering from some sort of hipster syndrome—you know, that affliction that keeps the cool cats from liking anything that's popular. I liked Super Lemon Haze, but why didn't I love it? Everybody else did.

It made me wonder what, exactly, makes a strain a Cup winner. I can't imagine three thousand stoners agreeing on much of anything, let alone deciding which pot is the best. It can't all be advertising and goodie bags. There has to be some intrinsic quality to the herb that makes it stand out in a crowd.

With the exception of the indica-dominant Vanilla Kush from Barney's that finished behind Super Lemon Haze in 2009, most of the top three finishers in the last five Cannabis Cups have been sativa or sativa-dominant varietals, usually with some Haze genetics, or at least the word "Haze" in the name.

In recent years flavors seem to be popular. Besides the citrus appeal of Super Lemon Haze, popular strains have had notes of vanilla or other essences; for example, the pungent minty sage of Arjan's Ultra Haze #1, the fruity spice of G-13 Haze, and the subtle chocolate scent of Chocolope.

But I wonder, seriously, if anyone could taste the nuance of flavor after sampling five or six or seventeen strains in a day. I've heard from some people that that's exactly why Super Lemon Haze keeps winning—because the flavor and high is able to cut through the other weed.

Or could it be some meteorological bias toward sativas? Maybe they tend to win because the Cup is held in November when the Amsterdam days are short, wet, and gray. Perhaps if the Cup were held in the summer and people could sit outside in the late-evening sunlight with a cold beer, they'd want that relaxed body-stone of an indica.

Of course, you can still get cold, wet, shitty weather in May in Amsterdam. The first two weeks I was there it was freezing cold and constantly drizzling. Everyone, and by that I mean everyone who's Dutch, apologized about the weather. I would walk into the local coffeehouse to buy a café latte and the barista would shake his head sadly and tell me how sorry he was about the rain. I ate one of the ever-present *broodjes* at a café and the waitress lamented my misfortune at being there during such awful weather. They say it as if the weather took some strange turn, like it's not always cold and wet, like that's not the norm.

But it was as cold in May as it was when I was there in November. Perhaps even colder. For sure it was wetter. And what comes between November and May? Winter.

One morning I bumped into my landlord. He gave me a grim look and said, "I'm sorry but it's only going to be seven degrees today." He seemed genuinely sad about it.

To be honest, at the time, I didn't mind the weather. In fact, I was happy it was cold. It kept the garbage from stinking. The fifth greenest city in Europe was heading into week two of a citywide garbage strike.

There were mountains of trash everywhere. Some piles were seven or eight feet high, blotting out views of the canals. In the Rembrandtplein—a large square lined with bars and bistros—they'd put up an eight-foot-high fence to hide the trash. Tourists, mistaking it for some kind of avant-garde art installation, would stop and pose for pictures in front of it. I was tempted to point out that it was just a giant pile of garbage behind a fence, but then I stopped myself. Who am I to say it isn't art? My landlord told me to put my garbage in a pile on the street with other people's garbage. The next day there was a sign on the trash that read: "Keep your garbage off the street until the strike ends." I know it was directed at me and I'm not being paranoid: It was in English.

The cold weather had an interesting side effect. It forced me to buy a scarf. In Amsterdam most everyone wears a scarf. Men, women, little kids. They wear scarves on their bikes, on the streets, on the public transport. They wear them in bars and restaurants. They probably wear them in their homes. The scarves are stylish, to be sure, but I don't think it's only a sartorial affectation; I think the Dutch wear scarves to keep their necks warm. They're very pragmatic that way.

What's funny to me is that with a scarf knotted around my neck, the perception of who I was suddenly changed. Before, when I entered an establishment, I was greeted in English, but now, with the scarf, I got lots of hearty greetings in Dutch. Strangers would nod and say *"Morgen"* or *"Middag."* And it wasn't just in the restaurants and bars. Dutch people would approach me on the street and start talking to me in that indecipherable, vowel-filled language of theirs. I'm almost tall enough to be Dutch and definitely pale enough, but the scarf sealed the deal. The scarf turned me into an Amsterdammer. And, in a practical Dutch kind of way, it kept my neck warm.

While the crappy weather might be one factor driving the dominance of sativa-winning strains, I think the real reason is Haze. The mysterious, mythological genetics of Haze add a kind of wild card to the experience of smoking. Some early seed catalogs warned that the effects of Haze were "not recommended for inexperienced smokers—too trippy, too profound" and that Haze was "known for an extreme, almost psychedelic spaciness."

Is Haze the key to dankness? Is it that simple?

The other thing that excites Cup judges—and honestly, pot

smokers around the world—is novelty. They like new tastes, new smells, new sensations. Right now there's a fad for strains that are purple. The Purps, Purple Urkel, Purple Kush, and other dark-colored strains are all the rage. As Franco says, "At the moment in California, anything that is Kush and purple is very popular, that's what's trending. So it's been trending here. I think that very soon there's going to be a comeback to true, true sativa flavors. California has been on the indica side for a long time and I think that's gonna change."

Of course, the purple-colored pot isn't necessarily genetic. You can shock a plant into turning purple by exposing it to the cold, icing the roots, and other botanical tricks. For a real breeder, the way to find something new—and maybe get an edge over the competition—is to go back to original landrace genetics and build something unique from that.

Arjan and Franco of Green House Seeds are producing a series of documentaries called *Strain Hunters*. Professionally filmed with production values that exceed most documentaries, the films follow Arjan, Franco, and a dreadlocked Australian with a somewhat philosophical bent named Simon as they tromp around third-world backwaters looking for unadulterated, un-hybridized strains of landrace cannabis. Although the films are geared toward people with an interest in cannabis, they are surprisingly entertaining and informative. Arjan and Franco are good-humored and intelligent hosts, and their obvious concern for the environment and the people struggling to eke out a living in a harsh world make the films more than just stoner movies. So far they've made *Strain Hunters Africa: Malawi Expedition, Strain Hunters: India Expedition,* and *Strain Hunters: Morocco Expedition.*

I asked Franco what they were looking for when they went on these trips. He typically gets a bit fervid when he talks about cannabis, but he was especially animated when discussing this project.

"When we go out, we are looking for a lot. The purpose of strain hunting is to find landraces, to find new cannabis genetics that have been isolated for a long time in various remote places on the planet and that are necessary for finding new cannabinoid profiles and new terpene profiles."

For those of us, like me, who haven't taken a botany class in, oh, ever, here's a brief definition: Terpenes are the essential oils and resins produced by plants. In the case of cannabis, terpenes give a strain its flavor. Cannabinoids are the chemical compounds that give the plant its various medicinal and psychoactive properties. There are, at least, sixty-six distinct cannabinoids found in the cannabis plant, and their concentrations, combinations, and the way they interact with receptors in the brain change from strain to strain.

"It's not only an activity that improves our gene pool for breeding new strains," said Franco, "but it also preserves these genetics for the future and prevents these plants from extinction. It's also an activity that may help find new medicines. At the moment no one is really organizing, collecting, or keeping these genetics, so we try to do that. We can see that every once in a while you need to refresh the gene pool—to keep making new strains, new flavors."

The films are also a brilliant marketing ploy, increasing brand loyalty by showing Green House's commitment to discovering and developing new strains of cannabis, while giving the armchair stoner an educational adventure.

But for all the effort these strain hunters put into their genetics and landraces and breeding programs, they all agree with Grey Area's Jon Foster. At the end of the day it's the plant that matters. If you don't grow it with love and care, it will never be dank.

I realized I needed to talk to some farmers.

Although Holland has a lot of great indoor cannabis growers, there really is no place like California when it comes to farming. It's blessed with rich soil and a perfect climate for growing things. The Golden State has more than twenty-five million acres of farmland and is the number one producer of plums, almonds, pistachios, lettuce, grapes, kiwifruit, asparagus, broccoli, celery, garlic, and just about every other fruit or vegetable you can name, including cannabis. According to Paul F. Starrs and Peter Goin's *Field Guide to California Agriculture,* published by the University of California Press; "Marijuana in California is big business, likely the largest value crop (by far) in the state's lineup, and it is perhaps the single largest commodity produced in California, including tourism."

California is where I would find some serious outlaw farmers.

Chapter Eleven

California Über Alles

It was a covert op. It had to be. It was dangerous enough that I was going to an illicit Mexican cartel grow site in the Sequoia National Park, but to go there with the biggest pot grower in Tulare County could attract the attention of both federal law enforcement and the Mexican mafia if we weren't careful. The DEA and La Eme are notoriously humorless organizations, and it was important to avoid any interaction with them if possible. The best approach would be a stealthy one. So we were going in commando style—which is not to say that I wasn't wearing underwear.

I rendezvoused with E, my contact's right-hand man, at a roadside café high up in the Sierras. He got in his car and I followed him, driving through the backcountry along deeply rutted dirt roads for miles, until we reached a small cabin nestled in the woods. The cabin was some kind of old hunting lodge that had been moved—wall by wall—from the national park to this secluded spot. A rotted out RV sat under an oak tree, the words "Hunting on a Budget" crudely spray painted on the side.

The interior of the cabin was rustic and cozy, well kept and comfortable, the walls lined with knotty pine and pictures of hunting scenes. An ashtray the size of an overturned Frisbee sat on the table overflowing with cigarette butts. Maybe it was a Frisbee.

E isn't much of a talker—at least not around people he's just met—and has a distinctly languid, somewhat bleary-eyed disposition; he doesn't seem bothered by much of anything, happy to take a seat on the threadbare sofa, stroke the rectangle of blond beard that juts off his chin, and wait for whatever. It's a countrified style of patience, and helps explain why he's such an avid duck hunter.

E turned on the TV to make sure his DVR was recording. He wasn't going to let our covert op make him miss his favorite shows. I watched as the menu flashed on the screen. He had about a dozen shows he recorded regularly. They were all duck hunting or deer hunting or fishing shows. Who knew there was so many hours of programming dedicated to annihilating wildlife?

But E had more than an enthusiast's interest in the shows.

"I wanna make a TV show."

"A hunting show?"

He nodded. *"Huntin' on a Budget."*

I pointed out the window, toward the dilapidated RV in the yard.

"With that?"

He lit a cigarette. "No fancy trailers, no ATV four-wheelers, we don't even use fancy guns. It's just . . ." He paused for dramatic effect. *"Huntin' on a Budget."*

I'll be the first to admit that I don't know anything about hunting shows or their demographics, but there's a certain white trash winsomeness to the idea that I thought might actually make good viewing. "You know, that just might work," I said.

E sat back and exhaled a plume of smoke into the middle of the room and cracked a smile. *"Huntin' on a Budget."*

He liked saying it.

And then his phone rang.

The big engine of the Chevy pickup growled as we drove up Highway 180, a windy two-lane that snakes through the Sierra Nevada foothills, up toward the top of the mountain. We couldn't risk parking the truck by the side of the road—it would be spotted for sure—so the plan called for a tactical insertion.

The mountain rose up on our left, broken by fingers of rough canyons, scrubby pine, and the occasional droopy-looking motel. On the right side of the road was a beautiful vista, rolling foothills of chaparral, and, in the far distance, California's Central Valley, millions of acres of orchards bearing olives, plums, pluots, and peaches stretching off into the horizon. It isn't called "America's Fruit Bowl" for nothing.

The view was magnificent because it wasn't impeded by any forest or mountain or hillside. The right side of the road was a sheer cliff that dropped off hundreds of feet into the valley below.

Hank Williams III, son of Hank Williams, Jr., and grandson of the country legend, sang his smartass hellbilly country punk on the truck's stereo—a song about drinking whiskey and smoking weed and fucking around with your best friend's girlfriend—as we pulled into a turnout and waited for the cars behind us to pass.

My guide on this excursion is considered the top cannabis grower in what is arguably the best cannabis growing area in California. He's friendly, intelligent, and complicated; a transcendental meditating libertarian, a perfectionist pot farmer, barbecue aficionado, and all-around family man named Crockett.

Crockett looked back at me, stroked his goatee, and cracked a grin. "Ready to do some bushwhackin'?"

I wasn't sure, exactly, what "bushwhacking" meant, but I thought I had a general idea. I flashed a thumbs-up. "I'm good to go."

The Guru, one of Crockett's business partners and close friends, made a tight face. "Just watch out for the poison oak."

The plan was for E to drop us off and then come back and pick us up in one hour exactly. We actually synchronized our watches.

E checked the rearview mirror, looking to see if any traffic was coming up the mountain behind us. "Looks clear."

Crockett gave a nod and E hit the gas, gravel spitting out from the truck's tires as we shot out onto the road. The sudden tire-squealing urgency signaled that we were now in commando mode, and everything needed to happen fast. For the first time I realized that they were actually serious about this covert "black op" stuff. Perhaps going to the Mexican cartel's grow site wasn't such a great idea after all.

E drove fast. The sheer drop on the right was dangerously

close to my window, giving me flashes of the valley floor followed by a bowel-churning surge of adrenaline. I wanted to tell him to take it easy, but it wouldn't have mattered. We were committed. It was go time and we were going.

The truck roared up the road about a quarter of a mile, into a long sweeping turn along the face of a sheer cliff. And then, without warning, E hit the brakes and the truck lurched to a stop in the middle of the highway. Crockett, the Guru, and I bailed out of the truck as fast as we could. E didn't wait around to see if we made it. As soon as we were out, he took off in a cloud of swirling dust.

I looked up and saw Crockett leap off the edge of the cliff and disappear. The Guru followed. I hesitated. I'm not the kind of person who likes to jump off a cliff when I can't see where I might land—I guess you could say I'm one of those "look before you leap" types—so I ran to the edge and took a quick peek. The drop was steep, maybe a good seven or eight feet down, but it wasn't totally vertical. I could see a landing zone where Crockett and the Guru were waiting. I followed their example and cannonballed off the precipice and into a dense thicket of manzanita, scrub pine, and what turned out to be poison oak.

I met Crockett and the Guru on my first trip to Amsterdam. We were on the same flight, we were all headed to the Cannabis Cup, and we struck up a conversation. They claimed to be two "construction workers" on an adventure, but I didn't believe that for a second.

I had ended up following Crockett through the kaleidoscopic mindfuck that greets passengers at Schipol, making our way down to passport control and baggage claim. Now I was following him as he dropped to his hands and knees and began to crawl through the dense brush like some kind of boot camp puke dragging his belly under coils of barbed wire.

Crockett is big, maybe six one or two and the kind of guy who is a lot stronger than he looks. When he smiles he radiates a friendly, laid-back charm, but when he drops his sunglasses over his eyes and tightens his jaw, he can suddenly look like a charter member of Sonny Barger's motorcycle club. He's partial to cargo shorts and flip-flops, but on this scramble down a cliff face he opted for sensible hiking shoes.

Crockett keeps his goatee scruffy and his long black hair in a ponytail, banded along its length into a thick whip. He has the look of a bona fide mountain man, which shouldn't be surprising: He grew up in the Sierras and spent his youth working for the park service. In fact, he'd considered a career as a ranger before he got caught up in the world of marijuana farming.

When he offered to take me to one of the Mexican cartel's illegal grow sites he told me to "bring the stuff you'd take on a camping trip." I'm not much of a camper. I don't really see the point in sleeping outside when there's a perfectly good hotel nearby. So when I head out into the world for some "nature," my survival kit is usually just a swimming suit, a credit card, and a corkscrew. But even if I was the camping type, I don't think I'd have the gear to go scampering down the face of an overgrown cliff or crawling on my belly through brambles and thickets of pungent brush. This was hard-core bushwhacking and, if you asked me, what we needed for this excursion was a machete. A sharp machete.

Crockett turned to me, as if suddenly remembering something important. "Oh, yeah. Be careful. It's been a bad year for snakes."

A really sharp machete and a gun.

I squatted in the brush, noticing that I had inadvertently crawled into a bank of poison oak, and heard the click of a lighter behind me. I looked back to see the Guru firing up a joint.

"You're smoking?"

The Guru shrugged. "I smoke more than most people."

He is also more allergic to poison oak than most people. Fortunately he has a lean and compact body that seemed to glide through the chaparral that Crockett and I crushed under foot. If anyone from the Mexican drug cartel was down below, they'd heard us. We sounded like a pair of rhinos crashing through the brush.

The grow site was supposed to be abandoned. It was busted a couple of years ago in a major operation—the DEA and local police had seized more than eight thousand plants—but you never know if the cartel has slipped back onto the site and started working it again. It's happened in the past.

Crockett raised his hand in some kind of commando signal and we stopped. I wasn't sure if he'd heard a rattlesnake or what. The brush was so thick I couldn't see more than a few feet ahead. Crockett pointed down the hill to his right. It took a moment for my eyes to adjust and then I saw it, through a thick screen of brambles. About twenty feet below us was a large depression carved into the side of the cliff lined with black plastic.

We shimmed through the tangle of brush toward it. The scrub and trees poked and snapped at me from all sides, drawing blood like arboreal piranhas. My backpack, which carried my notebook, camera, a bottle of water, and a few energy bars, had mesh netting on the side that kept getting snagged on the branches. I started, stopped, started, stopped, and got frustrated and yanked on the thing, causing branches to snap back at the Guru's face. He didn't seem to notice. He was preoccupied with not touching any poison oak, adopting the pose of a soldier trying to surrender, sidling through the woods with his hands raised in the air.

I reached Crockett and the pool dug into the hillside. It looked like a Jacuzzi Fred Flintstone might use, carved out of the rock to collect and store rainwater, and big enough for Fred, Wilma, and the Rubbles. A tangle of black irrigation hoses sprang out of it and crisscrossed the cliff. I wondered how this wasn't spotted by a helicopter and then looked up and realized that the entire hillside was under a dense canopy of trees.

We pushed on through the brambles, stopping when we came across evidence of civilization. A ragged pair of jeans hung from a tree. I guess there was no time to grab your laundry when the feds swooped in. There was a pile of trash: empty El Pato brand cans that had once held chipotle chilies, old cans of tomatoes and *frijoles refritos,* tortilla wrappers, and an empty bottle of Herradura tequila. The Guru found an old bottle of Caladryl, a lotion used to soothe the itching from poison oak, and laughed.

"I know how it feels, amigo. I know exactly how it feels."

Several bags of chemical fertilizer were piled up next to a large boulder and left to rot.

Further down we discovered more reservoirs and what must've been fifty miles of black hoses. In an area allegedly swarming with rattlesnakes, the tubing slithering on every square foot of the cliff made me slightly jumpier than I usually am when I crawl through snake-infested forests.

We entered a small clearing cut into the thicket. Crockett reached down and scooped up a handful of dirt. It looked different from the dust and rocks we'd been scrambling through. This looked like soil, like the kind of dirt you'd find in your vegetable garden.

He let it sift through his fingers.

"Here's where they planted some."

We walked further into the clearing and I sighed with relief. The ground here was only pitched at about forty-five degrees—unlike the near vertical slant of the rest of the cliff—and I was able to stand upright without gouging my head on overhanging manzanita branches. The Guru pointed to a spot on the tree trunks where branches had obviously been cut. He adopted a kind of wise-ass Colombo tone and said, "These marks appear to be man-made."

Someone had come through with a pruner and strategically sawed off branches allowing just enough sunlight through the canopy for the plants to grow without exposing the operation to snooping helicopters. Imagine pruning a national forest.

Crockett drew a map in the dirt, showing how the cartel had cultivated three large tiers that spread out across the face of the cliff. It was ingenious really; the scale of the operation and the stealth of the execution was mind-boggling. The Mexican cartel had taken an overgrown cliff in a wild forest and turned it into a vast, secret marijuana farm that was invisible from the road and cleverly camouflaged from the air.

In fact, this site was discovered by accident. The sweeping turn off U.S. 180 is only a quarter mile from a little roadhouse motel called the Snowline Lodge. One evening a couple of locals were sitting out on the front porch drinking beer when they smelled the distinct aroma of Mexican cooking drifting in the twilight breeze. I should note that, sadly, there isn't anything resembling a Mexican restaurant up on the mountain. Curious or, perhaps, hungry, the locals followed their noses down the road until they reached the edge of the cliff. They couldn't see anything through the canopy of forest, but they knew for sure that some *al pastor* was being grilled somewhere down below.

Crockett and the Guru refer to this part of the Sequoia National Forest as "the Battlefield." There are hundreds of acres of national parkland being used as covert grow sites for the Mexican cartel. For the cartel, it's easy. They drop a couple of *campesinos* and supplies off in the middle of nowhere and come back in a few months and collect the harvest. The cost of production is low and the profits are huge, as much as $100 million a season.

For the park rangers and local law enforcement, the job of catching the cartel is nearly impossible. The crops are invisible from the air, because the cannabis is planted to blend in with the natural growth pattern of the forest. The only sure way of finding a site is to send foot patrols into the woods, and even then, the brush is so thick that without a team of experienced bushwhackers, a ranger on patrol could walk right past a large operation and not see it.

Even though he's a licensed grower in California, providing connoisseur-quality weed to medical marijuana dispensaries, Crockett sees the cartel as a nuisance. It brings the scrutiny of federal law enforcement to the area. Although what Crockett does is legal in the Golden State, it remains a federal crime, putting him at risk. The cat-and-mouse game between the cartel and law enforcement causes disruptions, even occasional violence, and as someone who grew up in these mountains, Crockett has legitimate concerns about the environmental impact of this guerrilla farming.

"The Michoacáns rip up the hillsides. They leave trash. They cause accidental fires." He pointed to a pile of fertilizer next to a rock and asked, "What do you think's going to happen when those bags of fertilizer break open?"

"Michoacáns?"

He shrugged. "I don't know if that's really who they are,

but when they arrested some guys a few years ago they were all from Michoacán, so that's what we call them."

He seemed equally annoyed by the quality of the marijuana they're producing.

"The Michoacán shit's not bad, but it's not great. It's nothing like what they could grow here if they used good seeds."

I can't tell if that's his local pride coming through or just his professionalism. He really believes that if you're going to go to all the trouble to grow here, you should grow something amazing.

A quick check of the synchronized watch and it's time to head back for our rendezvous with the truck.

After scrambling up the cliff, where I got a face full of poison oak but, fortunately, didn't encounter a rattlesnake, we hunkered down, out of sight, by the edge of the road.

The Guru relit his joint and took a couple more hits. He exhaled and announced that he needed lunch.

We heard a low rumble and Crockett stood.

"That's her."

The pickup skidded to a stop a few feet above our heads and we scrambled up, back into the truck and back on the road.

Crockett didn't just come by his profession accidentally. He's a third-generation pot grower.

"I've been around it since I was a little kid. My mom's family and my dad's family."

"Your dad was a grower?"

Crockett shook his head. "My dad was a nine to five-er, worked in an office his whole life, but I had people on my mom's side."

He paused to light a cigarette. "I was always fascinated with it and it just kind of pulled me in."

"How old were you?"

"I started just helping out, working for bud. I was probably sixteen or seventeen and it evolved into growing it myself and getting help from a lot of people that I knew who were already in the business for twenty, maybe thirty years."

I was impressed. "That's some old-school pot farming."

Crockett nodded. "We figured that in my family, it started

around the late fifties, early sixties. So we've been growing for almost fifty years. In this area."

That, as I'll find out, is an important point for him.

"Did you always plan to be a grower?"

"It worked out to where I would go get a job doing construction or something like that and it would get to the point where the job was interfering with my growing so I would have to weigh it out and be like, what's going to make me more money? So then I started taking jobs that would allow me to grow. Like I worked up here in the national park as a cave guide for six or seven hours a day and I'd still have time in the middle of the day to tend my plants."

"It seems like you love growing weed."

Crockett took a drag on his cigarette and exhaled. "I do enjoy it. It's a lot of work, a lot of stress. I enjoy it 'cause I'm passionate about creating new things. I enjoy everything about it except for the law factor."

One of the provisions of California's medical marijuana law is that patients are allowed to grow six mature plants and twelve immature plants. But since most patients are either not farmers or don't have the room, they sign an affidavit allowing a professional cannabis farmer to grow their plants for them. This way, farmers like Crockett and the Guru can grow their crops legally, at least legally in California.

We turned off the main road and proceeded to roll through some beautiful countryside, down one canyon into a valley, up another. I couldn't fully appreciate the scenery as the constant rolling hills and switchbacks were making me a little queasy. Crockett pointed out places where the Mexican cartel were growing.

"They go in there, just off that creek."

A couple of switchbacks later the truck slowed down again.

"They're up there, somewhere. Maybe a couple miles in."

The Sequoia National Forest has a limited number of roads, and is one of the largest wilderness areas in the entire United States. Most of it is accessible only by hiking or bushwhacking in. With that kind of privacy and the optimal weather the area affords, it's no surprise that it's a popular place for illegal grow ops.

We drove down into the bottom of a canyon, into a little outpost of civilization nestled among a stand of tall pines, and stopped for lunch at a restaurant called the Pinehurst Lodge.

From the outside the restaurant looked like a big dusty barn with a battered, rusted-out sign above the door. Inside it looked like a big dusty barn, only with a bar along the far wall and battered, rusted-out tables and chairs scattered around the main area. Would you be surprised if I told you there were deer heads decorating the wall?

There was a small general store attached to the lodge, a shop that seemed to be stocked with two of everything: two cylinders of Comet, two boxes of Tide, two cans of Dinty Moore beef stew. It was the Noah's Ark of country mercantile and yet with the mold-stained concrete floor, chipped concrete walls, and ancient shelving, it looked a lot like the basement of my grandmother's house in Kansas City.

I studied the menu. Crockett offered some helpful advice. "Don't eat the fish."

I'm not sure I saw any fish on the menu.

E plugged some quarters into the jukebox and a plaintive country song came on. It was Brad Paisley singing the story of a man whose wife is making him choose between her and his love for fishing. E sang along with the chorus: "I'm gonna miss her . . ."

I turned to the Guru. "Do a lot of tourists come here?"

The Guru laughed. "If a tourist shows up here, they're lost."

There was only one person working the lunch rush—and by that I mean there was only one person to wait tables, work the bar, and cook the food—a well-built middle-aged woman who looked like she could hold her whiskey and throw a punch better than me. She smiled, gave me a wink, and warned us that she'd "try to do her best" with our order. I appreciated her honesty.

I have to say that the Sierra Nevada mountains are beautiful, stunningly so. It's no accident that they've been designated national parks and protected wilderness areas. But I was less convinced that they were prime locations for growing cannabis. The soil is rocky and it gets freezing cold at night. It was the opposite of the kind of environment you'd find in a greenhouse in Holland or an indoor grow room. As we sipped cold Pacificos and waited for the food, I turned to Crockett.

"So what makes this area so special for growing cannabis?"

Crockett suppressed a soft burp and considered my question. He didn't say anything for a minute as the Guru went outside to smoke a cigarette and E sauntered over to plug more money into the jukebox. I didn't know if they had left because they were bored or because Crockett was going to reveal some secret.

"We can grow our outdoor here from the end of April to the middle of November without any problem as long as we're below two thousand feet. Above that you've got to harvest by Halloween. That gives us a really long flowering time—not as long as some tropical areas but what you miss in the tropical areas are the cold nights and season change."

I'm not a farmer, but that didn't stop me from trying to sound like one.

"Aren't cold nights bad for the plants? I mean, isn't that why all those orchards have fans and flame throwers to blow warm air across the fields?"

Crockett shook his head.

"Temperature has a great deal to do with quality, bud size, and crystal production. It has a huge influence on everything. It influences some varieties more than others, but the ones it does influence, when the temperature drops you see a pickup in density, crystal production, coloring. I really like the cold nights and warm days. It really does a spectacular job on our outdoor. I can take our outdoor plants down to Los Angeles and they can't tell the difference."

He smiled. "Until I show them the indoor."

Like a classic French vintner, Crockett believes in the concept of *terroir,* the idea that the land and location where the plants live—the climate, soil, and peculiarities of geography—have a huge influence on the quality of the end product.

"You can tell if something's been grown in the mountains. You can taste it."

I was skeptical. You can taste the mountains? That sounded suspiciously like a beer commercial. "How's that?"

"Altitude has a lot to do with it. Right now we're at seven thousand feet and there's less oxygen and more CO_2, which is good for the plants. And we have higher amounts of UV rays, which produces more crystals because they're trying to protect the plant from the sun. Crystals are a whole protection deal—that's why they're there."

We were interrupted by the waitress/bartender/cook carrying an armload of cheeseburgers that turned out to be surprisingly tasty. The smell of the food brought E and the Guru back to the table.

As we ate, the Guru gave me detailed instructions on how to deal with poison oak. The minute I got back to the cabin, he said, I needed to shower in cold water with lots of soap. I looked at my arms. There was no sign of anything other than some scratches and dirt. The Guru chewed on his burger and shook his head.

"It takes a couple days before anything happens. When I get home, I'm showering with a soap they use on nuclear submarines."

"What kind of soap is that?"

"It's what they use if they've been exposed to radiation."

On the way to what they call "the airstrip" we stopped at another grower's house who I'll call Slim. He's Crockett's best friend from high school and part of a network of loosely affiliated California-legal growers in the area.

We pulled into the driveway to find Slim standing in the shade of a modern-looking farmhouse sharing a pipe with a dude who sported a thick reddish beard and was wearing a logo T-shirt for some alt rock band underneath a long-sleeved flannel shirt.

The Guru looked over at Crockett.

"Who's the dude?"

"He moved up here from Orange County to try and grow."

With a pair of cheap, oversized sunglasses and an ironic trucker cap, the OC grower looked like a poster boy for the Hipster Relocation Program.

By contrast, Slim is a lanky, energetic country boy who

seems to be constantly sunburned and delighted by everything. We got out of the truck and Slim offered Crockett a hit off the pipe.

"What is it?"

"Goo."

Crockett took a toke and passed the pipe to the Guru. The Guru took a hit and passed the pipe to E. E sucked down a long pull and passed the pipe to me. I declined. I didn't feel much like smoking. I'd just crawled through a hillside of poison oak, eaten a cheeseburger, and been advised to shower with soap used to wash nuclear waste off, so the idea of smoking a heavy indica like Goo didn't appeal to me at all. "Besides," I told myself, "you're working." I didn't know it at the time, but this ritualized circling of the pipe and lighter would continue for the rest of the day.

On the way over we'd been talking about dankness and what Crockett thought the word meant. Crockett picked up the thread of the conversation with Slim.

"What do you think 'dank' means?"

Slim reloaded the pipe and thought about it.

"I dunno. Good bud."

Crockett pointed to the pipe and asked, "Do you think this is dank?"

Slim looked at the pipe. "The Goo? Yeah. It's dank."

Crockett shook his head.

"That ain't dank."

The OC grower shifted his feet, looking suddenly uncomfortable. I'm guessing he grew the Goo and wasn't prepared for a spontaneous critique of his bud with a group of connoisseurs, especially if they weren't liking it so much. He checked the time on his iPhone, like he had an appointment or something, then looked over at his SUV.

Crockett held up his hand and revealed his criteria for dankness. "You gotta have all five of these to be dank." He counted them off on his fingers. "You got appearance—what I call bag appeal—the flavor, the smell, the taste, and the effect. Dank should have all five of them."

I started to point out that flavor and taste might actually be the same thing, but then I second-guessed myself. Maybe they're not. One could be the aftertaste. Sometimes the smoke tastes one way when you inhale but leaves a different taste when you exhale.

The Guru shook his head.

"No, man. Dank is a kind of super-stinky bud. You know, bad-smelling diesely skunky shit."

I disagreed. "Sweet-smelling bud can be dank."

Crockett nodded. "The five parts make good bud dank, and that helps you sell it. And that's the name of the game."

The OC grower, who'd said almost nothing the entire time we'd been standing there, suddenly took on a slightly aggrieved expression. He waved to Slim and skulked off. Crockett watched him go and shrugged.

"The Goo wasn't dank."

Slim had just returned from a family fishing trip in Alaska and was eager to show off his catch. He gave us a tour of his freezer, pulling out frozen filets of salmon and halibut, and excitedly recounted the details of each particular catch.

He held up one large hunk of halibut. "I didn't think I was gonna get this guy on the boat."

We jumped back into the truck and Slim piled in with us. He broke out a couple of fat nuggets of something he called Ghost OG, and while Crockett drove, pipe loads of the Ghost were passed around.

As the pipe and lighter made their way around the cab, Slim began to rhapsodize about Alaska. "They hunt wolves up there now. They used to be, like, protected, but they killed a lady jogger."

Crockett looked over. "Wolves attacked a jogger?"

"Yeah," Slim nodded, "so now they can hunt wolves again."

The Guru exhaled a plume of Ghost smoke out the window and passed the pipe. He looked at Slim. "Who jogs in Alaska?"

E flicked the lighter, ready to fire up the bowl, then asked, "Was she jogging in the woods?"

Slim flapped his arms in an exasperated shrug. "I don't know where she was jogging. That's not the point."

The Guru looked at me. "The jogger might not feel that way."

E stroked his beard philosophically, like he was in an imaginary episode of *Hunting on a Budget,* and said, "I'd be surprised to see wolves in the city."

Slim kept waving his hands around, trying to get the conversation back on track. He was stoned and his face flushed as he flapped his gangly arms in the air.

"Listen." He turned to make sure we were listening, then continued. "So in the gift shop of the lodge they got wolf pelts and wolf teeth and stuff like that. I thought about bringing some back for you guys, because they're so cool."

The Guru's face collapsed into a puzzled frown. "What would I do with a wolf pelt?"

Slim had already considered this. "You could make some sweet boots."

The conversation immediately derailed into an in-depth discussion of the pros and cons of using wolf fur for boots, but eventually, just like the pipe, it came back around to Slim, who'd just finished sucking down another hit of the Ghost OG.

"But then I saw this hat. You wouldn't believe it! It was fuckin' unbelievable. It's a hat, but like, it's the wolf's head." No one was following this line of thought, so Slim clarified: "They turned a wolf head into a hat. It was totally cool."

Crockett looked at him quizzically. "A wolf hat?"

"Yeah, you wear the fuckin' wolf head on *your* head. It's got eyes and its mouth is open and fangs showing and shit. The ear flaps are the wolf's paws."

Slim demonstrated, his hands making imaginary wolf paws run alongside his head.

"Like, little wolf legs, hanging down on the sides."

It got quiet in the truck for a moment as everyone tried to imagine a hat made from the head of an Alaskan wolf. Slim sat back and grinned. "I would've gotten one, but it was eight hundred bucks."

The truck rumbled along a series of back roads, the pavement narrow and rutted or disappearing altogether in some places, as we navigated a series of switchbacks and climbed to a higher elevation. Eventually we reached "the airstrip."

It was a large parcel of land on top of a mesa, a meadow bisected by a dirt road that ran about a quarter of a mile from a weather-worn farmhouse to a sheer cliff. The former owners of the property were weekend pilots who used this as an improvised airport. A sign that said "Road Ends" was posted at the edge of the cliff and was all that stood between an unsuspecting driver and oblivion. The drop was steep and the bottom rocky, but the view of the surrounding mountains and canyons was breathtaking. An old couch had been balanced on the edge for taking in the view, but a recent windstorm had tossed it into the abyss. I looked down and saw its battered Ethan Allen corpse shattered on the rocks below. The only sign of civilization that remained was an old toilet, perched majestically atop a granite boulder.

We doubled back to the farmhouse. As the truck pulled to a stop a gangly young man burst out of the house holding an old pie tin with a blob of black gunk on it.

"Hey! Who wants some heroin?"

This was Slim's younger brother, Red. Like Slim, he's excitable, but unlike Slim he's got a mischievous streak. Red likes to cause trouble.

"Fresh heroin!"

Crockett shot Red a look that froze him in his tracks. Red stood there, hesitating, unsure if he'd made some kind of faux pas. Crockett pointed at me. "Do you know who this is?"

Red looked at me, then back to Crockett. "No."

"You don't know who he is or what he does?"

Red's face suddenly flushed crimson and I realized how he got his name. "No."

"And you come out of the house announcing you've got heroin?"

Red looked at the ground, his face the color of a ripe tomato. He poked at the black gunk in the pie pan and laughed nervously. "It's not really heroin. It's hash."

The Guru declared that he needed one of Red's sodas and so we trooped inside the farmhouse. I was surprised by the interior.

I guess I don't know what I was expecting, but it wasn't rustic or country; it was totally suburban. There were two large mismatched couches, a few overstuffed chairs, and a shag carpet in some kind of wretched autumnal color. It was like a classic rec room from the 1970s, the slightly musty den where young men sat around on their granny's old furniture smoking weed and watching television—which was, I suspected, pretty much what went on here.

A box of shotgun shells and a shotgun rested on the fireplace mantle, but I didn't find that unusual out in the boonies. What did strike me as strange was the fat little quail chirping away in a cage. A handwritten sign identified him as "Topper." Topper seemed to like the company. He chirped and hopped around, making the little oddball plume on top of his head wiggle and bob.

I watched Topper as several more bowls of Ghost OG were consumed and a few brave souls tried Red's hash.

And then it was time to walk.

Red picked up the shotgun and took the lead. He kept the gun at the ready and by way of explanation said, "It's been a bad year for snakes."

I nodded. "I've heard that."

Apparently it was also a bad year for bears and mountain lions. Crockett pointed out a big pile of fresh shit just off the trail. "Bear."

I began to think that Red needed a bigger gun. I'm the first to admit that I don't know much about wildlife or survival. I'm the opposite of that *Survivorman* guy on TV who gets dropped into remote areas to live off the land. But I can, perhaps unlike the Special Forces–trained survival experts, drop into any restaurant in the world and find something to eat. I can read a menu, whatever the language, like a Native American tracker reads the woods. Maître'd's and sommeliers don't scare me. What does give me a moment of hesitation is strolling through rattlesnake-infested scrub for the second time in one day, only this time with the bonus of being stalked by apex predators.

"What happens if we see a bear?"

Crockett smiled. "There's only one thing you need to take with you to survive a bear or mountain lion attack."

I took an uneducated guess. "Pepper spray?"

Crockett shook his head. "Someone who's fatter and slower than you."

That got a big hoot of laughter from everyone. I discreetly checked out my companions. They were all younger than me but were wearing flip-flops or clunky boots. I could probably beat them in a sprint, especially if the sprint involved avoiding carnivorous predators. It helped that I hadn't had any of the Ghost OG.

I followed Red and Crockett into a large cluster of manzanitas and pines to find several tall and healthy-looking cannabis plants. Crockett introduced each of them. "That's Island Sweet Skunk. That's a Chocolope. Over there are a couple Super Lemon Hazes."

Even though Crockett has the legal right to grow cannabis in California, he still planted his crops in a discreet, natural pattern, so that they would be hard to detect from the helicopters that the DEA and Inter-Agency Task Force fly over the region. It's a strange position to be in—legal or illegal depending on who's looking—and not without risk. In one well-known case, a cannabis grower named Bryan Epis was sentenced to ten years—the "minimum mandatory" sentence—in a federal penitentiary for growing plants in his basement, even though he had the legal right to do it in California

We zigzagged through the brush, following deer paths until we found another cluster of vibrant green, healthy-looking plants springing up and swaying in the late-afternoon breeze. I noticed a complex web of irrigation hoses spreading out through the woods.

Several of the plants were a strain Crockett had developed himself, a cross between an OG Kush and Skunk Haze that he sells to dispensaries in Los Angeles.

"This is a variety that we created about ten years ago that I began to call 'Banana' because it has a faint banana smell, but because I had mistakenly taken a name that had already been used, we renamed it Private Reserve. But locally, we still call it Banana."

"You'd think people would be able to copyright names."

But then I realized that on a federal level it would be difficult to copyright something that's illegal.

Crockett thought about that for a moment, then shrugged. "I don't care what they call it. It doesn't matter to me. I know what it is."

I have smoked Private Reserve a couple of times and feel like I can say with some confidence that it has all the elements of dankness going for it. It's got a nice balance of the muscle-relaxing stone of indica and the soaring high of sativa, but what really takes it to the level of dankness is the influence of the Haze, that special genetic something that provides a trippy psychoactive effect. It's a strain that's right up there with the best I've tried, but outside of a few fans in Los Angeles and the locals in the mountains, it's unknown.

I was curious how Crockett ended up developing it.

"Where did it come from?"

Crockett lit a cigarette and checked one of the plants as he talked. "I'd been searching for a long time for a particular OG Kush cut, and when I found it, I grew it for a while. It's called the Ghost OG."

"What you guys were smoking."

Crockett nodded. I looked over and saw Slim trying to shoo a giant grasshopper off of one of the Island Sweet Skunk plants, flailing his arms around like he was doing some kind of stoner kung fu. Instead of just killing the bug, he tried to trap it in his hands and relocate it. But the grasshopper wasn't cooperating and kept jumping around. Slim couldn't hide his frustration.

"Come on, little guy. Let's find you a different plant. C'mon."

Crockett stood and moved to check another plant.

"I like the Ghost cut because of the nug size and the yield. It's not a huge yield, but for an OG it's good. I mixed it with a local skunk variety we've been growing around here for years, which is actually an ancestor of the Haze that everybody in Amsterdam is crazy over. It's a mix between those two, and the genetics that came out of it were just incredible. It's just a freak of nature."

"What made you think of crossing them?"

"I thought they would cross well because they're just such different spectrums. The OG is so indica and the Skunk Haze is so sativa. The Kush cut was probably eight or nine years old when I got it and the seeds from the Skunk Haze are from back in the early seventies."

The timeline is right. Haze seeds from Santa Cruz circa early 1970s could be the original Haze Brothers Haze. Crockett continued talking while he checked one of his irrigation hoses.

"I wanted to mix that Skunk Haze with something because I knew it was something that no one else had. It's really unique."

He smiled, unable to hide the fact that he's proud of his creation. "And now I've got something that no one else has."

He carefully stubbed his cigarette out in the dirt and turned to me. "I'd like to enter that sucker in the Cannabis Cup."

I think Private Reserve would do well in Holland, and I say so. "The competition's fierce, but I think you'd have a shot."

Crockett didn't agree that the competition is all that tough.

"When I was in Holland I'd go to these world-famous places and get world-famous weed." He made air quotes around the words "world famous" as he spoke. "I'd take it back to the hotel and, to be honest with you, the similarities in them all, the effect in them . . ."

He paused and tried to articulate his concerns.

"A lot of them were so cerebral, they're breeding out so much of the indica, that you don't even know you're stoned. You just feel confused. It seemed like the effects were so light, I felt, I don't know . . . I was looking for something more. I really wanted to get annihilated one night and I couldn't do it."

Private Reserve or Banana or whatever you want to call it had a fascinating pedigree—it sure tasted like it might be related to the Haze you get in Amsterdam—but I was curious if it truly was the direct descendant of the original Santa Cruz Haze that Neville and/or Sam the Skunkman took back to Holland.

Crockett smiled. "Wait till you meet Jerry."

. . .

We got back to the farmhouse just in time to see an old Volkswagen minivan come bouncing up the dirt road. The van squeaked to a stop—the hand brake ratcheting in that distinct VW way—and then the side door slid open and unleashed canine pandemonium. Five Australian shepherds came pouring out in a gray and white and black-spotted explosion of barks and yips.

Jerry creakily climbed out after them, carefully lowering his feet to the ground. He looked to be in his early seventies, although it was hard to tell for sure. A giant tangle of gray beard obscured his face and he wore oversized glasses with Coke bottle lenses that grew darker or lighter depending on the sunlight.

Jerry gave Crockett a friendly wave and walked toward us in a kind of happy-go-lucky old coot skank, like he had a reggae song rumbling around in his cranium. He didn't move like an old man, and he was definitely not about to spend the afternoon playing bingo down at the senior center. Jerry's a hippie, one of the originals. He's an acid-battered and deeply fried member of the tie-dyed tribe who still walks the walk and talks the talk. When he's not growing weed, he performs the most noble of all hippie occupations: He is a Volkswagen mechanic. He wears it proudly, too. His hands and face and clothes and hair were covered in grease and oil. The dogs swirled around me, banging their noses into my body, giving me the canine version of a cavity search. Jerry shook my hand warmly, leaving my fingers with a faint trace of motor oil. He sighed and sat down on the front step of the farmhouse and began polishing a piece of engine, something he called a "rocker bearing," with a filthy rag.

The dogs growled at one another and fought for attention. I was scratching the head of one and another came up and unleashed a menacing growl until the dog I was petting slunk into a submissive position. The growler then presented his head for me to pet. I fought the impulse to smack him. I've never liked bullies.

"What's with this guy?" I asked.

Jerry laughed. "Oh, he's an asshole. That's why I didn't get him fixed. I don't know why, but I love my asshole dog."

The asshole dog went over to Jerry and plopped down on the

dirt. The other male was quickly back by my leg, and I resumed scratching him behind his ears.

A breeze had kicked up from the valley and it was suddenly very pleasant to sit outside in the shade and shoot the shit. A pipe had been lit, too. Another round of Ghost OG was passed among the group as Crockett, the Guru, E, Slim, and Red sat around petting the dogs and listening to Jerry tell stories. Jerry enjoyed reminiscing. His eyes lit up when he talked about the good old days of California, a bygone era when the dope was plentiful, the sexual revolution was in full swing, and the music was psychedelic.

Jerry has always grown marijuana. He didn't always sell it. Mostly he grew it for himself and his friends, constantly keeping a couple of plants going in the backyard in Santa Cruz or Mendocino or wherever he found himself. The bag of Haze seeds from the 1970s that Crockett used to make his Private Reserve strain came from Jerry. According to the story, it was part of a load that was flown by a pilot Jerry was friendly with. The pilot would fly down to Mexico in a Cessna Piper Cub, going under the radar, and pick up a plane load of weed. His memories are foggy on the details, but Jerry thinks this particular pot was picked up in the city of Monterrey in northern Mexico. It was just good Mexican sativa with a little something special about it. That's all anyone knows about the origins of Haze.

I asked Jerry if they used the word "dank" back in the day. He shook his head. "Never heard that word. We just called good dope"—his voice trailed off—"we called it good dope. Good weed. We didn't really have a word for it." He cracked a smile and chuckled. "Just having some was good."

The long day of covert ops and cheeseburgers and unlimited amounts of Ghost OG was finally taking its toll. People were mellowing out. Heads drooped and even the constantly twitchy Slim was having trouble keeping his eyes open.

I hadn't smoked anything and the Guru took pity on me, rolling a joint of Jillybean, a strain developed by a female strain breeder named MzJill at TGA Subcool Seeds. Jillybean is a cross between a rare Pacific Northwest Orange Skunk clone and a strain called Space Queen. It's a nice mix of indica and sativa and isn't nearly as heavy as the Ghost OG. In fact, Jillybean is delicious, tasting of citrus and mango candy. It's sometimes prescribed for treatment of depression, and I have to admit that smoking a little took my mind off the strange burning sensation that was starting to flicker across my arms; the first warning signs of an impending poison-oak outbreak. I reminded myself not to scratch until I could get in a cold shower.

Jerry continued to smoke the Ghost OG and polish the rocker bearing as he described his rich and varied career in the drug trade. For a while he and a friend had had a tabbing machine and specialized in producing high-quality blotter LSD. He moved on to making PCP until an explosion blew the wall off the rented house they were using in the suburbs.

When he wasn't playing with home chemistry or giving tune-ups to Beetles and minibuses, he worked as a smuggler. He and a couple of his friends used to borrow a boat and sail loads of pot between Santa Cruz, Mendocino, and the San Francisco Bay. His smuggling career ended when, on a stormy day, in an ocean surging with large swells, they missed the channel markers and ran the boat into a sandbar. The boat split in two and sank before they even had a chance to untie the life vests from the railing where they'd been secured. Jerry and his companions treaded water in the surge, trying not to get sucked further out by the tide or bashed into an outcropping of rocks by the waves. They were eventually plucked out of the water by the U.S. Coast Guard. They were even interviewed by a local news crew after their daring rescue.

Jerry finished telling the story and shook his head with a wry chuckle.

"Oh, boy. That was a bad day to take acid."

An hour later I drove up the winding two-lane road toward Sequoia National Park where I'd rented a cabin for the night. The sun was starting to set and the tops of the massive trees were splashed with a glimmering golden light. It was truly one of the most beautiful landscapes I have ever seen.

The cabin itself was rustic, but not as rustic as some of the tent cabins that were available. I had electricity. I had a shower.

I wouldn't have to worry about mountain lions or bears, and thanks to the local market I had a couple of bottles of ice cold beer. Now this was a kind of camping I could get behind.

I followed the Guru's instructions and put my poison oak–contaminated clothes in a plastic bag, careful to avoid contact with the chairs or bed or anything else that I might sit on, because apparently the powerful oil called "urushiol" that rubs off the plant and causes extreme allergic reactions stays active on clothing until you run them through a washer. I felt like one of those bomb squad specialists in a film, moving extremely carefully, unsure whether to cut the blue or the red wire, not wanting to spread the toxin.

I drank a beer and felt a little better, then took an ice cold shower. Hot water allegedly opens your pores and allows the urushiol to penetrate your skin, so I stood there, engaged in extreme exfoliation, scrubbing my entire body with soap, hoping that the frigid mountain water would rinse away as much of the poison as possible. It was not nearly as fun as it sounds.

I got out of the shower and warmed myself with a second beer. I felt confident that I had removed every last bit of poison oak, or at least I had removed the skin that was exposed to the urushiol. A week later I learned, in painfully graphic terms, that I had not scrubbed nearly hard enough.

I went to the park's restaurant for dinner and found it full of French tourists. They sat staring at the menus as if they were completely incomprehensible. The menu itself was simple American food: steaks, pasta, chicken, and the house special, the aptly named "Sequoia Burger." I ordered a Cobb salad and another beer and watched as a French father, looking Euro-suave in white pants and matching white polo shirt, attempted to order for his family. He pointed at the menu and spoke as clearly and carefully as he could.

"We would like zis 'am-burg-AIR."

The waitress nodded like a happy poodle, her ponytail swishing in the air.

"The Sequoia Burger?"

The Frenchman looked suddenly perplexed, as if something had gone horribly wrong. He turned to his wife, equally suave and attractive in a navy blue dress, for help. She smiled at the waitress and attempted to order.

"The 'am-burg-ER."

I admired her attempt at deciphering the nuances of the English language. Was it hamburg-AIR or hamburg-ER? She looked at the waitress expectantly. The waitress held her pen next to her order pad, ready to write.

"The Sequoia Burger?"

The French couple exchanged a worried look and consulted each other in rapid French. I wondered if the waitress had some kind of passive-aggressive disorder. The Sequoia Burger was the only hamburger on the menu.

Chapter Twelve

Natural's Not in It

The next morning Crockett took me to see a giant sequoia called the General Grant. I'm not really sure what it means to name a massive, ancient, and awe-inspiring sequoia after the whiskey-soaked curmudgeon on the fifty-dollar bill, but it certainly is an impressive tree. It stood more than 270 feet tall, which is up there, but what was truly mind-blowing was the size of the trunk—it's 107 feet in diameter. It's as wide as a house and looks a lot like a cross between a magical tree in a Miyazaki film and a large booster rocket NASA might've built. The General Grant is just under two thousand years old, a relative teenager for a giant sequoia. Some of the other sequoias are thought to be more than three thousand years old. That means that around about the time the Roman Empire ruled the world, these trees were already well established in the Sierra Nevada.

I'd been thinking a lot about plants and what "dank" might mean. From everything I'd learned so far, it seemed like dankness was a kind of natural perfection, the peak expression of a plant's genetic destiny: a perfect peach or a tomato picked at the absolute moment of ripeness. Dankness could mean a flower at its fullest bloom—which for a flower would be its sexual peak—or maybe it could be a massive tree that has watched over the mountains for more than two thousand years.

Crockett and I sat at a picnic table under a grove of tall sugar pines. He smoked a cigarette while I watched a handicapped squirrel forage for food. The squirrel looked like he'd been run over by a car at some point in his life—perhaps during his careless teen years—and now his back legs didn't work. But the excess baggage didn't slow him down; he cheerfully dragged himself from trunk to table and back again, leaving skid marks instead of paw prints.

I jumped when I heard a loud crack. A pinecone the size of a bowling ball had just come crashing down, cratering into the dirt like a meteorite. I looked up at the trees and noticed that they were full of pinecones. These weren't the cute, decorative pinecones that you spray paint gold and arrange as a festive centerpiece for your holiday dinner. These were killer pinecones, the size of footballs, dangling on branches more than a hundred feet in the air.

"Should we move?"

Crockett laughed and stubbed his cigarette out in the dirt. "The big sugar pinecones, now they won't hurt so much 'cause they're opened up and dry, but in the winter, when they're big and green and the snow weighs them down, they send people to the hospital every year."

He looked up at the trees and grinned. "But, hey, this is where the biggest stuff in the world grows."

A loud squawk reverberated through the forest.

"We also have the world's biggest woodpeckers."

Like I said, he's a mountain man. But I wanted to get him back on topic. "So if I want to be a top-notch pot farmer, what's the first thing I need to know?"

Crockett lit another cigarette. "The first thing you gotta do is create or find superior genetics. That's the name of the game."

This is the opposite of what the botanists in Amsterdam told me; to guys like Franco and Aaron it's all about how you grow.

"I'd heard it was all about the care and feeding of the plant. You know? How you grow them is what's important."

Crockett nodded. "For sure, for sure. But I truly believe, and this is gonna flip every commercial grower out, that fertilizer is the least important thing in growing. The most important thing is environment. The environment minus medium."

I didn't know what he was talking about. "Environment?"

"Everything from air flow, temperature, and CO_2, to the atmosphere you create."

"You mean you play classical music to the plants?"

He laughed. "Hey, if that works for you, do it."

He decided he needed to show me, so we get in his truck and drove down the mountain to visit one of his indoor grow rooms.

Although Crockett is a big proponent of *terroir* and growing outdoors in the crisp mountain air, farming that way limits the growing season. And like with most things of high quality, the demand outstrips the supply, so he has built several state-of-the-art indoor rooms that allow him to harvest year-round.

I've seen a lot of indoor grow rooms, ranging from a small, one-light operation tucked into the closet of a studio apartment in Amsterdam to an industrial warehouse filled with thousands of plants. It doesn't matter how big or how small they are; they all look like something from a science fiction movie, like how cosmonauts might farm on a space station.

Maybe it's because the indoor lights give off an unnatural glow—which is strange because they're supposed to mimic the full spectrum of natural sunlight—or the fact that the walls, floors, and ceilings are often painted white with silver Mylar–covered vents and ducts hanging from the ceiling and dangling off walls.

It reminds me of *2001: A Space Odyssey*. You know, the scene where Dr. David Bowman floats through the guts of the machine to disconnect HAL, the renegade computer. I don't know why, but whenever I'm in a grow room, I expect to hear some patronizingly calm computer talk to me.

It seems unnatural that plants could grow so vigorously in such a sterile, controlled environment. But they do. They grow like crazy. By providing optimum nutrients and constant, full-spectrum lighting, indoor growers have succeeded in coaxing cannabis plants to their fullest expression. A typical indoor plant will have dense nuggets of buds frosted with the tiny resin-laden hairs that go by the botanical name of "trichomes" but are more commonly referred to as "crystal" because they are clear and somewhat shiny. Trichomes are almost pure THC, the active ingredient in cannabis. People are always talking about how the "new marijuana" is stronger than anything they ever smoked before, and that's due to the increased trichome production caused by indoor growing.

There's a delicious irony in this because it was the U.S. government's intensified "war on drugs" with its threat of "zero tolerance" and harsh "mandatory minimum" prison terms that caused farmers to move their crops inside. The DEA crackdown led to a boost in potency and a boom in demand for high-quality genetics.

It also created a whole new industry. Open an issue of *High Times, Weed World,* or *Skunk* and you'll find pages and pages of ads for indoor grow equipment: lights, nutrients, irrigation systems, filters, fans, DVD grow guides, and even special units designed to fit in your closet and provide everything you need to grow superior weed. A plant that thrives in almost any climate, in any country in the world, now mostly grows indoors.

As we drove down the mountain, I asked Crockett when he started growing indoors.

"I didn't get into indoor until '98, '99. I'd been around it but I didn't have the facility. Before 2000, outdoor was worth a lot more than indoor so I'd be able to make a killing on my plants and then snowbird—just work in the summer and play in the winter."

He smiled. "Until I got my family."

He was referring to his wife and two kids, his dogs, and his massive barbecue smoker.

Crockett listed off the stuff he used to build a successful indoor room. "I got one-thousand-watt high-pressure sodium lights, vented, A/C, carbon filters. I run CO_2. I've got controllers that control the environment and a super-insulated building."

"Where did you get all that stuff?"

"The equipment's basic stuff you can buy anywhere. I don't have any super special laser beams or anything."

The truck meandered through the iconic California scenery, passing rolling hills dotted with oak trees and chaparral, down into the flats of the valley with its farmhouses and fruit orchards, until we passed through a couple of metal gates and headed down a road made of hard-packed dirt. Crockett spun the wheel and the truck shot through a gap in a fence and plunged down a steep hill, where it skidded to a stop in a swirl of dust.

I looked over and saw a building, about the size of a double-wide trailer home, sitting in a small area that had been dug out of the side of a hill. It looked like one of those prefabricated offices that you see on construction sites. Only this building was windowless. This was the grow room.

I followed Crockett past some two-by-fours and scrap wood piles to the door. Crockett unlocked it and I followed him into a small workroom—kind of a foyer—a mixed-use business office, seed library, and laboratory. He stopped and lit a cigarette.

"My stuff is all organic, all from soil. And to be honest I don't fertilize a whole hell of a lot. A lot of the stuff I get around here is natural stuff. We know people who have turkey farms, chicken farms, stuff like that, or we get composted material that's all made here naturally. We have a greenhouse that's down in the farmland that's a huge complex that makes organic worm castings. We get truckloads of that. It's one of our major secrets in our growing."

"So there is a secret ingredient."

He laughed. "Go down to the fertilizer store and look at the hundred different bottles they got. Turn them around and look at what they're derived from. Most of them possess the same stuff. These big-time guys like Advanced Nutrients will tell you that if you add this, which is almost exactly this other stuff, at this certain time it's gonna do this for your plant. It's just all true bullshit. Okay?"

I nodded. *True bullshit.* I believed that.

Crockett leaned forward, continuing. "You can get everything you need by staging your soil, which means putting the stuff into the soil that you want your plants to uptake. You can get different types of additives that go in your soil that break down at different rates. Now an indoor garden, it's actually really hard to do this with. It's easy to add a bunch of shit to your fucking soil and say 'This made a big difference.' But to really know what happened is different because some things take longer to break down than others, and your pot is only in there for sixty days or whatever the flowering period is. So if that bat guano you put in there isn't partially soluble, then you're just throwing it away. It'll never break down in time for it to be used. So these companies that are selling you all this stuff are basically selling you stuff that you're throwing in the garbage can, and you're paying top dollar for it because you think it's growing your plants better. In actuality, just use good soil, good base fertilizer, a good secondary fertilizer, and it'll be fine."

"So staging the soil is the secret?"

"There's no real secret, yet there is: The secret is experience. That's the whole deal. You've got to stick with it. Don't get discouraged if you sink ten grand into a room and four grand into power and it all goes to shit. Because that happens. You're a farmer and farmers get one good crop out of four, so save your money."

Crockett opened the door to the grow room and flipped on the lights. I was hit by the smell of budding cannabis—it's a nice smell—and a blast of cold air from the air-conditioning units keeping the room cool. The lights stuttered and blinked to life revealing a clean white room brimming with plants.

Powerful grow lights hung from the ceiling, suspended by chains; a simple house fan circulated the air; drip irrigation hoses spidered out to every plant, and a CO_2 tank kept the carbon dioxide level high. It really did look like a space station.

I followed Crockett in and he walked me around the small room, pointing out the different varieties he was growing.

"That's Super Lemon Haze. This is interesting. It's L.A. Woman from DNA Genetics."

I looked closely at the L.A. Woman, a cross between L.A. Confidential and Martian Mean Green. The buds had a nice stink and the thick dark leaves indicated that it leaned heavily indica.

Crockett continued his tour. "That's some Cheese. And these are mostly Private Reserve."

He had about twenty-five or thirty plants flowering. In a smaller room off to the side was an equal number of plants growing in what's called the "vegetative stage"—the part of a plant's life when it's just growing and not worried about flowering. I guess you could say they're prepubescent.

Flowering in cannabis plants is triggered by the change in daylight hours. In the spring and summer, when there are long days and short nights, the plants grow big and lush. When fall comes and the days are suddenly shorter, the cannabis begins to flower. To mimic this natural rhythm indoors, plants in the vegetative stage are blasted with light eighteen to twenty-four hours a day until they reach a good height. Then they're moved into the main grow room and put on a schedule of twelve hours of sunlight and twelve hours of complete darkness. This switch in light hours makes the plants bloom.

A professional grower like Crockett keeps the vegetative and flowering rooms separate so he can maintain constant cycles of crops. With four or five indoor rooms running cycles like this, and counting his outdoor crops, he must grow a lot of weed.

"How much cannabis do you produce a year?"

Crockett thought about it. From the look on his face I thought that maybe he'd never actually bothered to add it up before. He finally came to a number but delivered it with a shrug. "Maybe two hundred to two hundred fifty pounds per year. But that's not all from me. I have people who work for me, and partners, and everything else that goes with that."

Before I could calculate in my head how much he was potentially grossing he read my mind. "A lot of it is put back into the gardens. To build our prototype garden like this costs about thirty thousand dollars and then another three thousand to run a cycle through it."

He stopped to take a drag off his cigarette, then continued the tour, pointing out various gizmos and doodads. It's important for indoor grow rooms to be clean—you don't want mold or insects that might attack your plants—but Crockett's room was cleaner than any I'd seen. It looked like an operating room in a hospital. The lights and the watering system were all automated and on timers.

"Do you even need to be here?"

He nodded. "Oh, yeah. When people say, 'Oh, this isn't doing so well' or 'This is getting pests,' I ask them, 'How long are you spending in your room?' You need to notice things. Does the room get hot at this time? Is this fan kicking in? You need to spend five maybe six hours a day in your room."

I can't say that I was surprised by his meticulous obsession with details. If I'd learned anything from Crockett, it's that the small stuff is what makes the difference between growing good weed and great weed.

"I see how you got your reputation."

He shrugged. "Reputation is a big thing. Around these parts I have a reputation for having the best marijuana. And when I go to other places like L.A. and I look at the quality of their herb—I've been to twenty or thirty shops—they all got really good herb, but the herb that I produce is right there if not superior. I've given a clone to some growers down there and they tried to grow it in a massive, twenty-light operation and it turned out okay. But it didn't have the resin production that I get."

"Why's that?"

"I think you have to grow in a manageable facility—for one person or two people. Once you start getting into massive facilities with multiple people with multiple ideas of how to grow, you get conflicting ways of doing things. Even if you have a specific style on what you think is the way to do it, you're telling this person what to do and they might have a different idea and they're doing it reluctantly because they believe something else."

He paused for a second, then continued.

"It's not to say that my way is the right way. It's just to say that if you're going to do it, be passionate about it. If you think you're doing it the right way, keep going and perfect your style. It's just like cooking or anything else. There's a million different ways to grow and they're all right. But the way I define a good grower is that he's happy with what he's producing. If you're growing a good garden and you're happy with it, you're doing it right. If you're not happy with your garden, if it's not producing enough or you're not getting the quality you want, then you might want to take some advice."

"But what if you create a system? Like a factory?"

He shook his head. "If you're doing it yourself and you know what you're looking for it's going to be better. If you're depending on a bunch of people to do it, it's going to be inferior, which is what the big boys do. You have to grow with love."

"With love?"

Crockett patted his chest. "Always with love."

It reminded me of what Jon Foster of Grey Area said: The best herb is handmade—artisanal cannabis grown by people who sweat the small stuff.

Chapter Thirteen

Organoleptic in Berkeley

If dankness is subjective—and it seems to me that it is—how can we define it? One person's awesomely *dank* might be another's totally *schwag*. Everyone's taste is different and, like the wise man said, there's no accounting for it. Some people prefer pinot noir to merlot, bourbon to scotch, lager over pilsner. I choose vanilla over chocolate every time, a fact that baffles my wife.

In the wine world—and let's be perfectly honest and say that no one is more snobby or pretentious than gastronomes and oenophiles—connoisseurs have developed a system for evaluating the relative merits of fermented grape juice. Robert Parker, the famous wine critic and editor of *The Wine Advocate,* uses a one-hundred-point rating system to define the relative dankness of wines. If he awards a vintage a high number—anything over ninety points is "outstanding"—the value of that wine skyrockets and previously unknown vintners become discovered. Some of them even become famous. It's kind of like *Star Search* for booze.

Parker went so far as to have his nose insured for a million dollars, a fact I find a little strange. Does he get the payout if his olfactory glands stop functioning or does his nose have to be lopped off in an accident? Or was it some kind of publicity stunt? Is he just hyping his nose? If I was going to insure one of my protuberant body parts for a million bucks, it wouldn't be my nose. But then my schnozzle is not the sensitive instrument that Parker claims to possess. I can't tell the difference between a ninety-two-point-rated 2004 and an eighty-eight-point 2005 in a vertical tasting. Perhaps it's because I don't like to spit the wine out after I taste it. What's the point of that? That's like Bill Clinton not inhaling.

At the end of the day, whether it's wine, food, or cannabis, anything that is judged by using taste, touch, or smell is going to be subjective. Scientists can tell us why something tastes a certain way, how it stimulates the circumvallate papillae, but they can't make us like it.

I wondered if it was possible to develop a similar point system for cannabis. That led me to Berkeley, California, where the people at the Berkeley Patients Group are wrestling with this very question. They've begun a project to try to identify the essential qualities of high-grade cannabis by using organoleptic criteria. Organoleptic assessments use the senses—sight, taste, touch, and smell—to determine the character and quality of a product. It's the exact same way Robert Parker judges wine, and the USDA meat and poultry inspectors ensure the food we eat is safe.

The I-5 freeway begins in Tijuana and ends in Vancouver, British Columbia, linking the entire west coast of North America with more than thirteen hundred miles of concrete. Sure, it sounds impressive, but the section connecting Los Angeles and San Francisco is a deadly dull drive. It's like drawing a straight line for five hours, the tedium occasionally punctuated by the stench of feed lots and giant billboards that warn "Buzzed Driving Is Drunk Driving."

I arrived in Berkeley ahead of my scheduled appointment so I went down the street to a little place called Caffe Trieste to have a coffee.

NorCal, as the locals like to call it, is not SoCal. The denizens of the Bay Area seem a little mossy compared to sun-dazed Angelenos. Perhaps the difference comes down to the fact that the NorCal folk are a forest people and the tribes of SoCal are beach and desert dwellers.

NorCal natives have a strange lingo, They use the word "hella" as in "that is *hella* awesome news" or "this sashimi is *hella* tasty" and their tree hugger earnestness always makes me a little suspicious, like I'm not sure they really are doing a citywide composting program—maybe they're faking it just to make people from the weird, plastic gasbag that is Los Angeles feel *hella* inferior. I wouldn't put it past them.

Caffe Trieste was lively, every table occupied, and yet I didn't see a single person with a laptop, iPhone, iPad, BlackBerry, Kindle, Nook, or any other electronic device. Outwardly intellectual and defiantly atavistic, the NorCal natives were reading books made out of paper.

What they lacked in electronic gizmos they made up for with the ubiquitous yoga mat. There was one at almost every table. Some of them even had special yoga mat–carrying cases. I will admit that the customers of the coffeeshop did, in fact, appear flexible, sprightly even. There were women with Pre-Raphaelite hairdos chatting over tea, young stay-at-home dads with infants in strollers, and a surprising number of older gray-haired men wearing silly hats. There was a septuagenarian sporting a newsboy cap, a stocky man in a checked sports car cap, an erudite gent in a straw boater, and a guy in one of those Australian outback hats where the brim snaps on the side. I assumed they were professors from the nearby university. There's nothing tenured academics like to do more than revel in their own eccentricities.

The Berkeley Patients Group is housed in a former pancake restaurant. It's a peculiar structure, with colossal floor-to-ceiling windows that sweep out in a large semicircle on the street side of the building. It's a style I call "early IHOP," and it gives the architecture a vaguely sci-fi vibe—the same aesthetic that gave us Googie drive-ins and cars with giant tail fins. With the addition of a chain-link fence topped with sharp looping coils of razor wire, high-tech surveillance cameras, and armed guards in the parking lot, the dispensary took on an ominous look, like a heavily fortified former pancake restaurant.

Once past a guard in the parking lot and another checkpoint at the front desk, I was met by David Stogner, a friendly and gregarious young man sporting cool glasses and a seemingly non-ironic blazer—sort of a hipster version of Mr. Rogers.

True to BPG's mission statement—"to provide the purest, most effective, and affordable medical cannabis along with integrated holistic health services"—Wednesday is free acupuncture day at BPG, and David introduced me to a couple of the acupuncturists who provided the treatments. Other days are devoted to cranial sacral therapy, massage, legal assistance, and a hospice program. All are provided free of charge. They even offer arts and crafts.

David smiled. "We try to offer fun activities for our patients." Just like Mr. Rogers, it's all about being a good neighbor—although, now that I think of it, if I had to weave a lanyard or make a macramé planter, a cannabis dispensary might be the best place to do it.

David and I were joined by Brad Senesac, the communications director—one of the few men I've ever met who can wear plaid pants and actually make them look cool—and Debby Goldsberry, the director of the operation. Brad has a scathing, sardonic sense of humor and is so energetic it wouldn't surprise me if he just started running in place; he's the perfect foil for Debby's easygoing charm.

Unlike the stereotypes of stoners and potheads often portrayed in the mainstream media, these three are all reassuringly professional—there's not a dreadlock or stitch of tie-dyed clothing in sight—and look as if they could just as easily be pediatricians or executives from a Silicon Valley startup. This is not to say that they haven't rocked a tie-dyed T-shirt in the past.

In the world of cannabis activism, Debby Goldsberry is a rock star. Tall, with sun-kissed good looks and a quick smile, she radiates a forthright, Midwestern wholesomeness. It's disarming, until she starts talking and I realize she's a whip-smart policy wonk who's not afraid to rattle off the legal intricacies and the zoning arcana of complex city and state ordinances. She's like a walking, talking encyclopedia of cannabis activism.

I imagine she has to stay on top of it since interpretations of California's medical marijuana law vary from city to city and county to county and are in a constant state of flux and revision. Local government decisions are often handed down quickly, with little forethought or debate, creating a fluid legal environment for dispensaries and individuals.

Debby doesn't appear to be fazed by any of it. She has a longtime activist's single-minded focus, and yet, her ability to speak like a politician on the campaign trail is offset by a sense of fun and a sincere, easy laugh. It's hard not to be smitten.

She was one of the founders of the Cannabis Action Network—a group committed to educating the public about cannabis and sensible and safe access for adult users—and the first activist to be voted a "Top CelebStoner" by CelebStoner.com, a popular website devoted to cannabis news and celebrity drug use. She responded to my mention of this accolade with an ironic "Woo hoo!" She also cofounded several cannabis industry nonprofits, including Americans for Safe Access and the Medical Cannabis Safety Council. She's a former board member of the Marijuana Policy Project and on the steering committee of the NORML Women's Alliance, she's been a VIP judge at the Cannabis Cup, and she was named "Freedom Fighter of the Year" in 2011 by *High Times*. Basically she's a supermodel for intelligent grassroots activism. To follow her career is to get a brief history of the marijuana legalization movement in the United States.

As with many activists, a traumatic incident lit the spark for her, and, like it does for a lot of people, the incident involved police brutality.

In the early 1980s, the third Wednesday of every April was designated as the day of the "smoke-in." On college campuses across the country, students would gather, listen to the Grateful Dead or Phish or whatever music was conducive to large gatherings of pot-smoking young people, and share a mutual appreciation for marijuana. It was a peaceful, mellow, flower-power kind of event and was tolerated by most colleges and universities.

But when Ronald Reagan was elected president, he brought his arrogant new sheriff swagger and an intolerance for anything smelling faintly of disobedience or patchouli to the war on drugs.

Debby Goldsberry was there when Reagan's opening salvo was fired. A swarm of riot police attacked a peaceful smoke-in on the University of Illinois campus. The crackdown was bloody and violent, and left her stunned. "I was shocked," she said. "As a thinking person, I couldn't understand how something so peaceful and nice could turn into this."

Maybe another person would've run and hid, stopped

wearing peasant blouses, and renounced smoking pot, but Debby decided to do something about it. She and a group of like-minded students got organized, studied their legal rights, and learned how to protect themselves. The next year they held a successful smoke-in despite a campus ban and the threat of arrest.

Emboldened by their success, they expanded their activism and successfully organized smoke-ins on five campuses in Illinois. The year after that, colleges in Missouri, Iowa, Wisconsin, and Indiana held smoke-ins.

She teamed up with fellow activists Ben Masel and Jack Herer to create a "hemp tour," moving from state to state, town to town, campus to campus, speaking out, engaging, and educating local activists on their legal rights. They were trying, as she says, "to change the dynamic in the war on drugs."

In 1989, she formed the Cannabis Action Network—a group that continued the work she began with the hemp tour—expanding their efforts to include educating the general public on the beneficial uses of cannabis and advocating responsible use. After nearly a decade at CAN, and with California voters approving a medical marijuana initiative, Debby founded the nonprofit Berkeley Patients Group and was able to put her vision of compassionate and responsible cannabis activism into practice.

Unlike the moldy dogs and stink-eyed Kush dealers you find in some of the more disreputable dispensaries in Los Angeles, the BPG operates with a real concern for their patients, their patients' rights, and how they can improve their communities. They work hard to be good neighbors and a prime example of what a cannabis dispensary should look like. In other words, I could take my mother there and she would think it was nice and clean and professional. Maybe that's one of the reasons the feds leave them alone.

As Debby said, "We could get rich, but then we'd get busted, and that's not our mission."

The Berkeley City Council called the BPG a "national model" and declared an official Berkeley Patients Group Day to celebrate the dispensary's tenth anniversary.

Other cannabis activists, most notably John Sinclair and Jack Herer, have been honored by having strains of marijuana named after them. I think it's only a matter of time before Debby gets a strain named after her.

"How about Goldsberry Berry Kush?"

She laughed. "Oh. I don't know about that." Then she considered it for a moment. "Maybe if I got to choose the strain. I think I'd like that."

Brad, David, and Debby led me through the crowded lounge where I finally spotted a pothead cliché: a long-haired dude in a tie-dyed T-shirt taking a bong hit. We walked past a snack bar and a rack of small plants for sale—part of the DIY credo at work—and into the dispensary room.

Dozens of people sat in folding chairs, waiting patiently for their turn with the budtenders. There was an impressive selection of cannabis on display, each strain nestled in a glass case, presented like a gem in a jewelry store. The budtenders were friendly but maintained a serious, professional demeanor with their patients. They treated the cannabis as if it were a sacred sacrament or, perhaps more accurately, they treated the bud like it was medicine. I mentioned that it reminded me of some of the upscale coffeeshops in Amsterdam.

Brad cocked an eyebrow.

"If I can say it, I think we're better than a lot of the coffeeshops in Amsterdam. Everything is handled and packaged with real care. We're not just dumping it into some old plastic tub that hasn't been cleaned in a year."

On an average day, between four and five hundred patients come through the doors of the dispensary, and I'm not so sure the attention to hygiene is what accounts for the BPG's popularity. It could have more to do with the quality and variety of the cannabis on display.

The menu was projected on the wall behind the counter, and I watched the familiar brand names of high-grade cannabis scroll past. Brad turned to me.

"I don't know why they have to use some of these names. White Widow? Green Crack? Can you imagine a patient who might be new to cannabis, maybe a little nervous about it, coming in here and buying Green Crack?"

"Not really. No."

"We call AK-47 'Aff Goo,'" he said.

Aff Goo? I wasn't sure that was an improvement. It sounded like something you'd use to clean an oven with. "I like the name AK-47," I said.

"It's a good name for a band," David chimed in.

Brad continued. "The name thing is a real problem. We're working with growers to try and get them to change the names they use."

Debby nodded.

"Whenever there's a natural disaster the next batch of cannabis gets stuck with that name," she said. "I can't wait to see what Tsunami will be like."

I laughed, but Brad was serious. He pointed to the menu. "White Widow?" He shook his head, and I couldn't tell if he was disgusted or just deeply disappointed. "White Widow? Who comes up with that?"

I was surprised by the vocabulary that Debby, Brad, and David used. Not once did I hear the words "pot" or "weed." The plant was called "medicine" or "cannabis." Customers were referred to as "patients" or "members," and no one got high: They "medicated." Only once did I hear anyone say the word "marijuana" and that was connected to "medical." It struck me as a little odd, this kind of rigorous relabeling, but I understand what they're trying to do. They are making a determined effort to change the language, to alter the discourse. Debby is still trying to "change the dynamic"; if the language changes, maybe the culture changes, too. "Weed" becomes "cannabis," "stoners" become "patients," "getting high" becomes "medicating," and what was once an underground and illegal enterprise on the outskirts of society shakes off its youthful rebellion and counterculture ways, grows up, and joins the mainstream. Maybe it even gets legalized.

The Berkeley Patients Group has developed a form called a "bioassay sheet" that they use to evaluate strains of cannabis.

These same surveys were used by the Temple Dragons and celebrity judges at the 2009 *High Times* Cannabis Cup in Amsterdam to judge the seed cup categories. The forms look at all aspects of a bud of cannabis, taking into account the appearance and density of the flowers as well as the odor, taste, effect, and duration of effect. They are also interested in identifying flavors and scents common to cannabis as a way of systematizing and defining strains. They use descriptors not unlike the kind used to describe wine; words on their flavor wheel include "spicy," "earthy," "savory," "tropical," "citrus," "pine," "diesel," "fruity," "creamy," and "skunk." The forms also track the effects of a strain, ranging from sedative to euphoric.

After a strain is evaluated with a bioassay sheet, it's rated on their menu with a star system. The stars not only indicate the potency of the medicine but also the quality and purity of the plant and the way the plant was grown. But who awards these stars? Is there a Robert Parker–type weed sniffer in the house?

I was taken upstairs, through a succession of touchpad locked doors, to a lab on the second floor, the domain of self-described "canna-nerd," Eli Scislowicz.

Amid the NorCal hipsters and stylish professionals that administer the BPG, Eli looks a little scruffy, with the telltale slouch of someone who has played a lot of video games, but once he starts talking about cannabis, his mouth cracks into a sly smile and his passion for the plant comes pouring out.

"Eli was the right person at the right time for what we're trying to do," Debby Goldsberry said. "We're lucky to have him."

Eli is the intake specialist for the BPG. His lab is clean and scrupulously organized. This isn't about sorting seeds and stems at your kitchen table; this is serious business, the first hurdle a grower has to pass to get his or her product into the dispensary downstairs.

It's not necessarily an antagonistic relationship. BPG is actively working with growers to produce better quality medicine. When growers bring their cannabis into his lab, Eli sits down and examines the plant with them.

"This way they can learn how to be better growers."

When a crop is delivered, Eli examines it using what he calls the Herbal Medicine Intake and Evaluation Form (HMIEF). He hands me a copy. It looks like something you'd see a doctor using in an emergency room. It's similar to the bioassay forms but is more specific to medicinal uses and follows California Health and Safety Codes 11362.5 and 11362.7.

He starts by giving the product a visual examination.

"Medicine that is past its prime will have a yellow to brown tinge and will have lost much of its terpenes to the atmosphere. As a connoisseur, I am not concerned at all with the structure or density of the medicine. It all looks the same once you grind it up."

Terpenes, as I understand it, are the chemical compounds that provide the various flavors. More than that, according to Eli, they often have beneficial uses. For example, a pine-scented strain such as Jack Herer can act as a bronchodilator. I can imagine that it might be useful, yet I can't quite see asthma patients firing up a spliff instead of reaching for an inhaler. The terpenes responsible for the citrusy taste of Sour Diesel also act as antimutagenic agents—they prevent cells from mutating—and could help fight various forms of cancer.

"Do you have a chemistry degree?" I asked.

He smiled and gave a sheepish shrug.

"My dad's a physicist at Caltech."

Eli then looks for evidence of mold, mildew, pests, or anything else that shouldn't be on the plant. What he can't see with the naked eye he puts under the microscope.

Next he checks for moisture content. Too much moisture and the smoke will be harsh; if the plant is too dry, all the terpenes evaporate. To make sure the moisture content is just right, Eli performs a "snap test." If the bud doesn't snap off, it's too wet, and if the snap produces a lot of "shake"—brittle trichomes—it's too dry.

Even though he's got a computer-driven microscope and various other scientific gizmos, he can tell a lot by giving the cannabis an inspection that would make Robert Parker proud. In other words, he sniffs it.

"The most important factor in my mind is the taste and smell. Not only does the flavor of cannabis make or break the enjoyability of the medicine, it may also play a role in the effect of cannabis."

This makes sense to me. What's the first thing I do when I pour a glass of wine? I sniff it. Perhaps nostrils are the starting point to try to determine dankness. "So you give the bud a sniff. What're you smelling for?"

"I look for an intense and clean flavor."

I kind of understood what he meant by "clean"—like how you don't want sushi that smells fishy—but he elaborated.

"By 'clean' I mean free of molds, pesticides, mildews, or other contaminants. All good medicine should contain no discernible trace of any of these."

"You can smell all that?"

He nodded. "Sure."

He went on to explain that contaminants also show up when you smoke, in the form of bad flavors and cough-inducing harshness. A harsh toke is often a contaminated toke. *And I thought it was just me.*

I realized that this process is similar to the one Jon Foster uses to determine the quality of cannabis he sells in his coffee-shop. But the BPG takes it a step further, preparing cannabis samples for laboratory testing for THC levels and to look for contaminants that organoleptic examinations might not detect. All of this is done to provide safe medicine for their patients.

But there's something else happening here, as well. By building a library of data on various strains, the BPG is laying the groundwork for fast-tracking FDA approval when cannabis becomes legal.

I sat down on a rolling stool, the kind you'd find in a doctor's examination room, and asked Eli if there's one strain of cannabis that's the dankest of the dank.

"I think that is an extremely difficult question to answer since we have not mapped out all, or even most, of the strains in existence based on the cannabinoid, terpenoid, and gene profile," he said. "It's like asking a jury to render a verdict after only hearing the opening statements."

He gave his scruffy facial hair a thoughtful scratch and

continued. "And, really, the best medicine obviously depends on the hands that have cared for it."

I admire BPG's commitment to quality and safety and the efforts they're making to be an asset to their community. For sure, a lot of dispensaries could learn from what's going on there. And I can't help but respect the trajectory of Debby Goldsberry's career, taking her activism for the plant and for social justice to new levels, and converting the belief that the weed should be freed to an engaged compassion for her community. But what about dankness? Can all this testing and organoleptic inspection find dankness?

"So say the bud makes it through all these tests and you confirm that it's good, clean cannabis—which is great, don't get me wrong—but what about some kind of baseline standard for dankness?"

Eli considered my question and then answered. "I personally want cannabis that is at least sixteen percent total THC. This is what I call 'potent medicine.' From an organoleptic point of view, this is medicine that has a heavy coating of resin."

Debby agreed with Eli. "I don't care if it's indica or sativa; I just like it strong," she said.

In the months since my initial visit to the Berkeley Patients Group, things have changed. For one they are growing, with more than seventy employees serving almost a thousand patients a day. With that rapid growth comes policies and guidelines to keep the service and cannabis quality high. BPG employees now wear uniforms, which in and of itself is no big deal; they are, after all, trying to reassure a nervous public, and having budtenders dressed in matching polo shirts lends a calming Starbucks vibe to the place. They also, interestingly, sponsor a Berkeley Free Clinic Truck and have their logo on a NASCAR race car.

Although she remains on the board of directors, Debby stepped down as managing director of the dispensary in early 2011 and has partnered with an urban planner and a real estate developer who specializes in nonprofit and affordable housing developments to create the United Cannabis Collective, a company whose mission statement declares them to be "dedicated to social equity, economic vitality, and environmental stewardship" while providing "medical cannabis, in all of its varied forms, and essential life services that improve the health, housing, and safety of all collective members." She is, as ever, committed to ending cannabis prohibition. And if the battle for social justice and cannabis freedom didn't keep her busy enough, she's also writing a book about the history of the anti-prohibition movement.

"It just seems like the next logical step for a stoner like me," she said.

Chapter Fourteen

He Blinded Me
with Science

I had heard the rumors—the rumble of gossip emanating from various members of the cannabis cognoscenti in Los Angeles. They were all talking about a low-key collective that specialized in rare and unique strains of connoisseur-quality cannabis. It was called the Cornerstone Research Collective, but it wasn't listed on any of the "weed finder" websites; they didn't advertise in *Kush Magazine* or *West Coast Cannabis* or anywhere else. It was hard to find and even harder to get into once you found it. The collective didn't accept walk-ins. They didn't solicit patients. You needed to know someone who was a member who was willing to vouch for you. It made me think of a secret society, like Skull and Bones at Yale or one of those obscure religious cults that pop up in Catholicism with frightening regularity.

If the rumors were true, once you were admitted, you would find the finest herb in all of Southern California. It was ganja El Dorado, a veritable temple of dankness.

I'll admit that, at first, I found this exclusivity annoying. One of the things I hate about living in proximity to Hollywood is the harsh and arbitrary judgment of the bouncer, the tyranny of the velvet rope. The people allowed to pass are often richer, more famous, and more beautiful than mere mortals. I know, that's the point of the rope—but it's dehumanizing and creates a bogus celebrity. It's all about the superficial flash and sizzle, the heat of the moment. VIP rooms are like man-made ponds stocked with manufactured "reality" stars and other artificially plumped, processed, and packaged performers. Anything wild or raw or real is excluded; the celebrities allowed to pass are merely laminated versions of humans, their expressions frozen by Botox, imperfections concealed by spray tans, and personalities replaced by fake tits.

I realize, of course, that the collective isn't the VIP room at the Colony or Sky Bar; I won't see glossy nipple slips of the budtenders who work there or gushing accounts of the growers who provide the weed, because they're not part of the paparazzi culture. The Cornerstone Research Collective flies under the radar; they've chosen to operate on the down-low. And given the fact that complaints from neighborhood groups about the flashy, in-your-face dispensaries have caused the Los Angeles city attorney to initiate a wave of closures, maybe the collective has shown admirable foresight. Maybe a little discretion is the key to survival in a tumultuous political environment.

I had sent the collective several emails introducing myself, but I had never gotten a reply. This wasn't necessarily an unusual response in the cannabis industry. Most of the time I had to find someone who knew someone who had a friend who knew a guy who might make an introduction. It was never easy, but what made it especially frustrating—or especially ironic—was the fact that I lived three blocks away. I passed the building several times a day and I was desperate to know what they were doing in there. What rare treasures were concealed within this impregnable fortress of dankness?

And then, out of the blue, I got an email from Michael Backes, the director of Cornerstone. It wasn't a response to my earlier queries. It was fan mail. He'd read my novel *Baked*—the story of an underground botanist from Los Angeles and his adventures in Amsterdam—and really liked it. I wrote back immediately, telling him that I'd heard great things about his collective and asking if it would be possible to come see for myself. He said yes.

The Cornerstone Research Collective is housed in a small, nondescript building on a characterless street in a relatively obscure corner of northeast Los Angeles. There's no signage—no green crosses or flashing marijuana leaves twisted out of neon. The only indication that there's even a business in the building is a small plaque on the wall that says "Cornerstone" and a generic plastic doorbell.

I punched the doorbell with my thumb and waited. By now I was used to the gauntlet of security when entering a dispensary: the gruff ex-marine answering the door, a shuffle through a metal detector, and then waiting to be buzzed through a door into a cage, to be buzzed into the dispensary. It is not unlike entering a minimum security prison. So I was surprised when a tall and elegant African American man, smartly dressed in a suit and tie, opened the door, flashed a friendly smile, and showed me in. Of course, for all I knew he was a Secret Service–trained ninja with an Uzi strapped behind his back, but he seemed more like a gracious host. I felt instantly at ease.

The interior of the collective looked like a dentist's office—simple and clean with comfortable chairs and a hipster soundtrack bubbling up from the iPod on the coffee table. As the young woman behind the glass checked my doctor's recommendation, I sat down and studied the menu. It was not like any menu I'd seen at any dispensary in California.

For starters, the list was small: fifteen strains of cannabis and four different types of hash. There were also a couple of edible items, a blood orange caramel that sounded tasty and something called a "double strength" cookie. While most dispensaries don't carry many of what I would call true sativas—they usually just offer Sour Diesel or Jack Herer—this menu leaned toward exotic landrace sativas and Haze. There was Thai Haze, an heirloom combination of tropical Thai sativa and Haze; Kilimanjaro, a pure African landrace sativa; a strain they called Nano, which was Maui Haze crossed with Island Sweet Skunk; an Original OG Kush, and several powerful indicas, including Sour Bubble and Afghani #1. I was instantly reminded of Jon Foster's meticulously curated menu at the Grey Area in Amsterdam.

After a short wait, I was invited into the dispensary room, where I met a tall man with a perfectly shaved head and stylish eyeglasses. This was Michael Backes, avid reader and the brains behind the collective.

Michael stood behind the counter in a light blue button-down

shirt and jeans. He is an outwardly intelligent individual, looking more like an avant-garde architect or Superman's archnem-
esis than your typical budtender. He seemed intimidatingly
brainiac, but that was defused by his quick smile and knowing
chuckle. And once he started giving me the tour of his menu—
opening the jars like a magician performing some amazing
sleight of hand—he couldn't contain himself; his enthusiasm
was irrepressible.

He jammed a glass container under my nose. "Check this
out. Thai Haze from one of the best growers in California."

I was hit by a deeply pungent and fresh aroma.

There was a freestanding magnifying glass craning over a
dozen jars on the counter. I held the bud under the lens and saw
a forest of crystals. Michael opened more jars.

"Woodhead. It's Grapefruit crossed with White Widow."
And another. "Habañero Haze. Blackberry crossed with Super
Silver Haze."

More jars were opened.

"Pineapple. Big Sur Holy Weed. Ultra Violet."

I sniffed some more and studied the crystal content and bud
structure under the magnifying glass.

"Here's a sativa from Hunan China."

Michael had an unrestrained, rapid-fire way of talking, and
he shuffled the large glass jars around the counter like a three-
card monte player. It didn't take long before I realized that he
might know more about cannabis than anyone I'd met. My
head was spinning.

He broke off a chunk of the Thai Haze and put it in a plastic jar.

"Here. I've got a couple things I'd like you to try."

He held up a small container with what looked like a cookie
crumb from a snickerdoodle.

"This is our C3 8-Ton dry-sift hash. It's insanely strong."

I didn't know what any of that meant, but I remembered
how much I enjoyed Aaron's dry-sift Sleestak hash, so I happily
agreed to sample it. Michael put it in a small paper bag along
with the Thai Haze and samples of Golden Pineapple, Kilimanjaro, and Nano.

He looked at me and asked, "What do you know about Delta 8 THC?"

"Nothing," which was true.

He opened a small refrigerator behind him and pulled out a plastic vial that looked like one of those things you get perfume samples in when you go to a department store.

"This is pure Delta 8 THC."

The vial was about a third full of some blackish gunk that reminded me of the stuff that gets stuck to your fingers whenever you try to repair something on your car. Michael explained that normally, when you smoke cannabis, you only get the Delta 9 THC, so the fact that someone had isolated this obscure cannabinoid is a real rarity. The Delta 8 THC came complete with instructions for dosage—"two drops per dose, buffered in edible oil"—and a helpful copy of the gas chromatograph scan detailing the chemical properties of the oil just so I could take a closer look at what it was. Not that I know the first thing about reading a gas chromatograph scan.

Michael started to hand me the paper bag and then held up a finger. "Wait a sec."

He went into the back room and came out with a machine-rolled joint.

"I've been playing around with this. It's a Cambodian sativa. Takes twenty-two weeks to flower and the grower wants nine thousand dollars a pound for it."

That's more than double the price a grower would charge for super-high-end bud.

He dropped the joint into the bag with the other samples and said, "Welcome to Cornerstone."

I took my goodie bag home and began, over the course of a week, to sample the various strains Michael had given me. I loved the Thai Haze. It had all the qualities I look for in a great sativa: a clear, uplifting high with a relaxing, clean, and energizing effect. It was like Ritalin for adults, the kind of mind-focusing pot you'd smoke if you needed to clean your kitchen, plant a garden, read a book, or build a bird feeder. It would probably be good if you were going to a disco, too.

I had been looking forward to tasting the Nano—after all I loved the Sweet Skunk I'd smoked with Franco in Amsterdam—but the Nano gave me a headache. Perhaps it was my mood.

I had better luck with the C3 8-Ton dry-sift hash. It was potent, like a good wallop of whiskey, but the effects were very clean; you got baked, but without the jumpy edge or paranoia.

I had threatened my wife with the idea of making a nice balsamic vinaigrette with the Delta 8 THC oil—a couple of drops in the mix should do the trick—but I ended up putting a drop in a glass pipe and smoking it. The gunk bubbled and vaporized; I inhaled. And nothing happened. I waited for a while and still, no effect. So I went about my day. About forty-five minutes later I realized that I was actually high—I just hadn't noticed. The effects of the Delta 8 THC were extremely subtle. I'll admit that, once I recognized the high, I enjoyed it.

Everything Michael had given me was as close to dank as I'd found. They were all potent and interesting—what I call "dynamic varietals"—that had been expertly grown and cured. Whether it was the Thai Haze or the Kilimanjaro, the smoke was smooth. I never coughed or tasted any funky chemicals or latent nutrients. Every single one of these strains was as good as you could get, and yet I wasn't completely sold. Was dankness ephemeral? Maybe Jon Foster was right—maybe it depended less on the strain and more on the situation.

I finally got around to trying the machine-rolled joint of pure Cambodian sativa. The Eagle Rock Music Festival is an annual event where the police close about six blocks of a major street in my neighborhood and up-and-coming local bands perform on a number of stages scattered around. There are folkish groups, salsa bands, DJs, mariachis, and local favorites Dengue Fever, Great Northern, and Wreck of the Zephyr.

I stood on the street and shared the pricey sativa with my wife and our friend Trevor. The joint itself didn't have much of a scent. There was no punch or pungency to it, so I have to admit that I wasn't expecting much. We passed the joint around—joints are meant to be shared—and my wife, who hardly ever smokes, gave up after two hits. Trevor and I continued smoking as we strolled down the residential streets until he said he'd had enough. I wasn't about to waste a nine-thousand-dollar-a-pound once-in-a-blue-moon sativa so I finished it.

It was a warm October night, typical of Los Angeles, and the sun was beginning to set, casting the world in violet and pink. In the distance I heard a band called the Submarines playing a kind of buoyant pop music. As the puffy clouds in the sky burned orange on their edges, then faded into deep purple, I began to notice the kind of visual pattern-recognition distortion and color enhancement that is typically the harbinger of early onset psilocybin. This tropical sativa from Southeast Asia provided a gleamingly clear, profoundly trippy high. I was completely baked, and yet never once was I too high to engage with the world. In fact, instead of the feeling of overwhelming paranoia that some people get when they're profoundly stoned, the Cambodian sativa made me positively loquacious. And, even better, it lasted for hours. This was as close to dank as I'd found so far and yet, I wasn't sure. Suddenly I began to question my perception. Would I be too stoned to recognize dankness when I found it?

My wife was quick to point out that, typically, the one strain that I thought might be closing in on dankness was virtually impossible to find and unbelievably expensive.

There's an aesthetic at work in places like Cornerstone and Grey Area and it springs from the personalities of the individuals running those places. Michael Backes and Jon Foster might not know each other, but they are definitely kindred spirits in their search for the best possible ganja.

I was curious about what drives them. What makes them go to such lengths to ensure that the quality of the cannabis at their stores is unique and uniformly world-class? They are, after all, the final point before the product reaches the consumer. The strain hunters and botanists have all done their work, the farmers have grown and cured the plants to the best of their ability, and then, if it's very, very good, the cannabis ends up at a dispensary like the Cornerstone Research Collective.

Michael agreed to meet me for breakfast at restaurant called

Auntie Em's, a favorite hangout of Eagle Rock locals, so while I guzzled hot coffee and nibbled on an orange-cashew scone, he worked his way through a bowl of granola and talked to me about his vision for the collective. I asked him if he'd ever been to Grey Area in Amsterdam. He nodded vigorously.

"Sure. Many times."

"I think you guys have a lot in common. Like at Grey Area, the choices on your menu aren't random, they're curated. What do you look for?"

Michael didn't hesitate. "Pharmaceutical-quality marijuana. That's what I'm looking for. And it's really hard to find. You find it by accident. You don't find it by a lot of intention."

"What? You stumble across it on your way to work?"

He chuckled and scooped up a spoonful of granola. He didn't eat it right away; he paused, trying to figure out the best way to say what he wanted to say. I watched as a lone blueberry balanced on top of the spoon.

"The reason is, is that pharmaceutical grade marijuana . . ."

The blueberry teetered for a second before he popped the spoon in his mouth and chewed. He washed it down with some coffee and switched gears, waving the empty spoon in the air to make his point.

"What we have now is a high THC marijuana, and that's all interesting and good, but with 430 ingredients in marijuana, there are ones that are much more interesting than THC. I mean at some level I think that some of the intoxication of marijuana I look at as just a side effect. Anything that impairs me, I view as a side effect. Okay? I want to get rid of the side effects."

"You mean you don't like getting stoned? Like what happens when you smoke an indica?"

Michael put his spoon down and looked at me.

"They're *all* indicas. Karl Hillig at Indiana University did a genetic study of ninety different landraces from around the world, the largest study of landrace cannabis ever done. One thing he found, very clearly in the genetics, is that all cannabis are indicas."

I must've looked confused because he gave me the kind of look that a sympathetic tutor might give a special needs student.

"There are three subspecies that we primarily use— subspecies indica—those are the narrow leaf tropical drug strains we call sativas."

I hate it when scientists go around renaming things. Everyone knows these plants as sativas, so I don't really see the point in changing their name. Do they think it makes them look smarter? Isn't it already confusing enough? It's like when paleontologists decided the brontosaurus was really an apatosaurus. Was that really helpful?

"Why don't we just keep calling them sativas? What's the problem with that?" I asked.

"The problem is that the guy who named cannabis was French. It was the eighteenth century. He never got on a boat; he never went to India. He was basing his taxonomy on drawings and specimens that, in a pre-refrigeration world, had either been pressed and brought back or were just, you know, rotting. And . . . he got it wrong."

Michael held his hands up and shrugged in a gesture that said "What's the point of blaming an old French botanist for such a colossal fuck-up?"

"Now we're starting to get it right, and the reason it's important to understand that all drug strains are indicas is that the three different primary subspecies seem to have radically different chemotypes."

I'm not a scientist. I haven't suddenly decided to change the names of things, and so at the risk of revealing that I had no idea what he was talking about I asked, "What are chemotypes?"

"Chemotypes means different ratios of the chemicals responsible for the effects. And I'm interested in dialing in the effects. I think the future of cannabis, the world's quote 'best' cannabis, will happen—it's not happening yet—when we have a better understanding of how about sixty different chemicals within the plant interact. Now in mathematics they have a thing called a 'combinatorial explosion,' which you get into when you've got too many variables and you start to examine the interactions of those variables."

Just the mention of mathematics made my eyes glaze over momentarily, and I briefly considered resorting to nodding my head encouragingly and letting the digital recorder run. Michael paused to see if I was keeping up. I nodded encouragingly and he continued.

"You're talking about a very, very complex endeavor to start to understand how cannabis works. And these kind of broad strokes are of limited value to patients. We've noticed that in our collective. We used to think 'Oh, we can just simply give you an indica or give you a sativa and get the effect that you're looking for.' It's finer than that. And the reason is that the *kafiristanicas*—the strains from Nepal that have the diesel- or fuel-like smell—they're the wild card. They're often called stimulating sativas. They're speedy, like a cross between cannabis and caffeine, and they totally change the equation."

"Is that because of the cannabinoids or the THC or what?"

"It's not THC. THC is a mild stimulant, but there are essential oils in cannabis that can really jack that stimulation, probably through some kind of synergy or interaction. But go ask anybody, 'How's this work?' and they don't have the answers. Probably the closest is Raphael Mechoulam's group in Israel. They're doing the most study on the more obscure cannabinoids, and now they're starting to study some of the other essential oils—starting to get some sense of the interaction."

Michael took a bite of his granola before continuing. "But it's early days. That's something that's wild to think about—that after all this time we haven't advanced a lot."

"Well, the plant has been repressed," I said, which is true. One of the biggest problems with cannabis prohibition is the U.S. government's ban on human research. Here is a plant that offers so much in terms of medical application and yet thorough scientific examination is illegal.

"But also we've gone down some dead ends," Michael countered. "First they thought that CBD [cannabidiol] was psychoactive. Then they thought it wasn't psychoactive. Now they think it modulates the psychoactivity. In other words, they don't know. And that's just one of the major cannabinoids. You've got CBC [cannabichromene], CBG [cannabigerol], CBN [cannabinol]—there's a lot of these. It's early days."

Michael took a sip of coffee.

"But what's cool, what got me interested, is that I wanted to figure out if there could be more predictability so that we could develop a breeding program that wasn't 'Wow, this gets me really stoned' crossed with 'Wow, this gets me really stoned.'"

"Not that there's anything wrong with that."

"No, there's not. But there's more to it. People chronically overmedicate with cannabis."

"They get baked."

"The reason they get baked is because it comes from a time of scarcity. It's like binge drinking. Right? I can't get it when I want it, so holy shit I'm gonna take in as much as I possibly can until I can't take anymore. What's interesting is the UC San Diego pain study on cannabis showed that there was something else going on with cannabis that we should pay attention to, even if we're using it as an intoxicant. Which is: There's a sweet spot of dosage. So when you find that there's a sweet spot of dosage for pain, couldn't there be a sweet spot of dosage for inebriation?"

I nodded.

"So wouldn't it be nice if we could do the same thing with marijuana and dose precisely? The tough part with alcohol is hitting that sweet spot and keeping it there, but I think that's easier with cannabis."

I don't mean to brag, but I can hit the sweet spot with alcohol easily. It takes exactly two margaritas. But, I'll admit that it took me years of trial and error to hone that kind of precision dosage.

Michael continued. "Say you want to write but you're a little anxious. You want to rid yourself of the anxiety but you don't want to interfere with your short-term memory or your recall in general, so what strain do you pick? You might have some ideas. A light sativa, something like that. What if you could really just dial it in? Like literally? You knew that this strain did exactly what you wanted. That's tough right now. It's a crap shoot, and I want to change that."

Michael is an enthusiastic speaker. Words tumble out of his mouth like machine gun bursts and I can see that sometimes his brain gets ahead of his mouth, causing his words and thoughts to overlap and collide. It's at these moments where he stops, midsentence, and reframes what he's trying to say.

"As a rule, most of the super-high-THC meds, you just go past the sweet spot so fast you're on the other side before you know it. The thing about it is, you know, we've got Everclear and 151 out there but we don't often choose them as our drink of choice. Okay? And really, effectively, some of these marijuanas now that are getting up in the midtwenties in terms of THC percentage, it's really hard to dose them."

"One hit and you're baked."

Michael smiled. "The problem we're suffering from is the fact that a lot of the marijuana we have access to, its cultivation is driven strictly by prohibition, therefore it's easy to grow indoors while a lot of the stuff that's most interesting are these tropical strains."

I think about Michael's menu at Cornerstone. Thai Haze, Cambodian sativa, Kilimanjaro: These are all tropical sativas, and they're varietals that you just don't see on other dispensary menus.

"Is it just because the tropical strains grow so tall? You can't really grow a fourteen-foot plant in your closet."

Michael chuckled. "They're taller than that. Some of them get to be twenty feet high."

That would make for an awkward houseplant. And then there's the flowering time. I can't imagine a professional marijuana grower like Crockett waiting six months for a single plant to flower, although the Cambodian was an amazing strain, and earning nine thousand dollars a pound might make up for the hassle and patience required to grow it.

Michael continued. "It's the difference between Two Buck Chuck and Petrus. 'Money is no object' cultivation just isn't popular yet, except for hobbyists and snobs. But that will change."

"I hope you're right. That Cambodian sativa was as close to dank as I've ever found."

Michael smiled a knowing smile.

"It's amazing. But nobody will grow it."

. . .

Although he was a biology major at Indiana University, Michael isn't a scientist by profession, but when he's not running the collective he gets paid to spend his time thinking like a scientist.

He has been a creative consultant on several movies, including *Iron Man* and the Spider-Man series, collaborating with the writers, directors, and visual effects crews to create the fictional technologies used by the characters in these films. The trick, and it's not as easy as you might think, is to take something as absurd as a scientist getting metallic arms permanently fused to his body, and make the science and technology that explains the phenomenon completely believable. Doc Ock's cephalopodic mayhem, Tony Stark's cold fusion pacemaker, and the DNA-concocted dinosaurs in *Jurassic Park* have all been graced by Michael's science-centric imagination. I'm not surprised to learn that his specialty is creating the scientific technologies used by Spider-Man's enemies.

It's not just superhero movies that rely on his expertise. He also provided technical and scientific consulting for author Michael Crichton on his books *Jurassic Park, The Lost World,* and *Timeline.* That partnership led to him cowriting the Sean Connery film *Rising Sun* with Crichton and earning an associate producer credit on the mutant-apes-gone-wild film *Congo.*

Apparently equatorial sativas can make you very, very smart.

"Were you always a connoisseur of pot?"

Michael laughed. "I was the kind of guy in high school and college who would walk around with a dopp kit—remember those?—filled with little tubes of different kinds of marijuana."

It was easy to imagine a young Michael walking the streets of Tucson, with a dopp kit full of samples tucked under his arm. But like me—and a lot of people—as he got older, he took a break from regular cannabis use.

"It's funny. I didn't smoke for twenty years. After college I stopped, got married, had a kid, and didn't smoke for a long time. But then I came back into it."

"Because the product had improved so much?"

Michael shook his head. "I found I was suffering from migraines and I didn't like Imitrex and those drugs and I found that cannabis acted as a prophylactic."

"Do you mostly smoke sativas?"

He nodded as he concentrated on getting the last few blueberries into his spoon.

"I specialize in sativas. We usually have a really good selection. We're getting some Nano today, and it's my favorite everyday sativa. The Nano is what I consider the current state of the art as far as perfect mood elevation with minimum impairment. It's giggle weed, but you're not sloppy. And that's what I want: I want precision and great laughs."

I watched him eat the berries and decided not to tell him that the Nano gave me a headache. "Why do you think Californians prefer Kush? Is it just because they haven't been exposed to Dutch genetics?" I asked.

Michael shoved the empty bowl off to the side and leaned forward.

"The light psychedelic sativas are going to be the future of marijuana. They're great for posttraumatic stress. They're great for anxiety and depression. What we need is cannabis to usurp some of the pharmaceutical inroads in treatment because the thing about it is, why should we take drugs at the loss of our affect? I think Xanax and a lot of these drugs cost you your affect. Also they make you less risk averse. There's this whole theory about the financial crash being caused by antidepressants because with antidepressants, one of the side effects is you lose your ability to assess risk."

"Wait. You're saying that the entire worldwide economic collapse was due to the fact that too many brokers and bankers and Wall Street traders were popping Klonopin and Xanax?"

Michael nodded.

"Talk about drug abuse . . ."

One of the things that impressed me about the cannabis I'd sampled from the Cornerstone Collective was how well grown and cured the herb was. This, as you can imagine, is a real problem for many dispensaries. A lot of growers are forced to operate illegally, and not all are as conscientious as Crockett and the Guru in the Sierras. There have been complaints of bud sprayed with household pesticides to kill bugs, and heavy fungicides are sometimes used to control mildew and mold. Finding scrupulous growers is a big deal.

"How did you find your growers? Do they have to pass some kind of test?"

Michael leaned back in his chair and stretched.

"It's been a process of about three years of cultivating patient cultivators. I have a very, very small group of cultivators within our collective who are on it. What we do is choose a range of chemovars, of chemotypes of cannabis, that have specific essential oil and cannabinoid profiles that have been tested at Steep Hill, and I just make sure we have a wide range."

Steep Hill Lab is a state-certified independent laboratory in Oakland that tests cannabis samples for purity, THC levels, and cannabinoid profiles.

"For a while you couldn't find good OG Kush. It was really rare. And you'd have these people come in and go 'I'm growing OG' and I'd say, 'Technically you're growing it, but you're not growing it well.' But that's changing. Now people are showing up with bags and you open them up and go 'Oh, wow.' They've really got it dialed in. It just takes a lot of work and a lot of time. It's interesting because it doesn't really scale. The bigger these grows get, the more the quality suffers. It seems that you need to really baby these plants. And it's that aesthetic difference again between Two Buck Chuck and a wine like Petrus or Opus One or whatever the benchmark for really good wine is, because the best-quality marijuana can't be scaled. It's going to remain small."

This seems to be a recurring theme. "The best weed is artisanal," I said.

"Exactly."

And it goes beyond just growing the plants. For Michael, curing is just as critical to producing connoisseur-quality cannabis.

"You look at tobacco and it's really interesting because the curing of tobacco is extremely well understood. The curing of marijuana is absolutely not well understood. In fact, tobacco has an advantage because it is poisonous, so a lot of things can't grow on it while it's curing. That's not the case with marijuana."

Curing is the final, and often most misunderstood, part of

processing pot. In the *Marijuana Growers Handbook,* noted cannabis expert Ed Rosenthal devotes a large chapter to mani-curing and curing buds. "Buds that are dried too quickly, with-out curing, retain more chlorophyll, which gives the smoke a 'greener' minty taste and rougher smoke, and often less intense odor." It takes a few days for the harvested buds to metabolize the chlorophyll in their cells but the result, a better tasting, bet-ter smoking product, is worth it. Not that it's easy. Curing is about finding a delicate balance between temperature, humid-ity, and air. Too much heat and the buds dry out and lose their flavor; too much humidity and you get mold.

"All the dangerous pathogenic molds are all cure problems," said Michael. "You'll hear about powdery mildew being a prob-lem for cannabis, but you can smoke a bowl of powdery mildew and it wouldn't do a damn thing to you. It's not a pathogen. It's not toxic. But aspergillus, fusarium, and penicillium are cur-ing problems that are shockingly common on cannabis because people don't know how to cure safely, or well for that matter. If there's one thing that I've brought to Cornerstone, it's well-cured cannabis."

Like Lex Luthor, Dr. Otto Gunther Octavius, and other misunderstood scientists, Michael has plans that go beyond sim-ply sourcing the best possible cannabis. He wants to change the way we think about growing and breeding the plant.

"What is really wild about cannabis is how fast it adapts. And that's what I want to do. I want to get more control over the en-vironment so that I can basically create a phytotron, which is an enclosed, completely computer-controlled environment. I want to control for more than just temperature and humidity. Here's an interesting thing about plant physiology: Plants eat light."

I admit I hadn't really thought about plants as light eaters, but I could see what he was saying.

"Plants are a little more advanced than we think because they can take light and effectively convert it into energy. And the systems by which plants respond to light is really interesting. There are these things called 'signal transduction pathways' in the plant, and if you hit the plant with a particular frequency of light, a gene will begin to express. What's cool is, it's kind of the control panel for the plant and if you start to learn what frequency of light to hit the plant with you can tell the plant to do whatever the plant is capable of doing and maybe a little beyond. And that's going to change this game forever. Because if you walk into a plant physiologist's office and say, 'Hey, explain to me the mechanism of flowering in plants,' they'll go, 'Well, there's a notional enzyme in plants.' You'll say, 'What do you mean, notional?' 'Well, we know that the plant is using an enzyme to flower. We just can't find it. We call it florigen, but we haven't really found it.' They don't know what's going on."

Michael leaned forward conspiratorially, as if someone at one of the nearby tables might overhear him. "There's some guys in Japan that say they've found it, but . . ."

He made a face that indicated he was skeptical of the Japanese botanists.

"What's the difference if they find this enzyme or not? Does it matter?" I asked.

"What it means is that we can hit the plant with specific frequencies of light. A few of these frequencies are beginning to be understood, but apparently many frequencies have yet to be determined. It's possible that there are frequencies that stimulate cannabinoid production to the genetically determined maximum for a particular strain. Or not. We don't know what the switch is, but that's what I want to find out. And that just changes the game."

I couldn't imagine your average pot farmer suddenly adjusting light frequencies to get notional enzymes to change the chemical composition of the plant.

"Farmers don't have the capacity to do that."

"An enlightened farmer does."

"What's an enlightened farmer?"

"For me, an enlightened farmer is somebody who knows how to use a liquid chromatography and PCR machine."

"What's a PCR machine?" It sounds like something he fabricated for Tony Stark's lab.

"Polymerase chain reaction. It's a DNA amplification device. So you can take a little DNA and make a lot more DNA and then find out what it is and then sequence it. It's basically chromatography for genetics. It really does allow you to do amazing stuff. I mean right now we're at 'Well, that plant looked good' and 'That plant smoked good' so let's put a bit of pollen on it. And that's cool. That's how a lot of great agricultural strains are developed."

"Like the pluot."

"Exactly. So it's not to knock the farmer. I just think the farmer can be supplemented by some science and it'll make the farmer's job easier. Look, by accident we bred CBD out of marijuana. Big mistake. Next week we're getting some cuts of a high-CBD, low-THC plant from a breeder in Santa Cruz, and I'm just going to open source it. I'm going to give cuts to every breeder I know in L.A. I just want people to start playing with these high-CBD strains and breed them into their OG Kushes. Because you'll get a much more interesting effect, for sure. The first time you smoke a really high-CBD strain, it's shocking how different it is. It's a really interesting hit. It's like our experiments with Delta 8 THC at Cornerstone. We're the only people who have Delta 8 because it's found in such minuscule amounts that you never get exposed to enough of it to be a significant dose. So it's fun to provide it to people and see what their reaction is."

"Do you really think this manipulation of the plant, by enlightened farmers or whomever, is really going to happen?"

Michael tapped the table with his knuckles.

"I think the next five years will really show us what we can do with cannabis. The LED lighting thing has just now started, and the LED revolution will be a real revolution in how cannabis is cultivated indoors. There are definitely certain advantages to indoor cultivation, but the environmental footprint is so horrifying that we've got to go to a more energy-efficient model. The carbon footprint is insane. And they churn out a lot of effluent, too. What do they do with their dirty water? They pour it down the drain. And a lot of that is filled with very, very high nitrogen and phosphorus loads. My line for a while has been cannabis is not a houseplant. If you want a houseplant get a ficus. I think greenhouses will be the future."

"Do you really think you can override a plant's genetics by controlling the light rays?"

"A lot of different strain developers make a lot of different claims. I'd like to see some of these guys prove it. You know? The claims of 'I've got the Congolese Green from the Burunga range,' you know? I don't know. Prove it, pal. We're going to start a project called 'Who's Your Daddy?' And what it is, is just tracing mitochondrial DNA from plants to establish a true taxonomy. Because, really, there's bullshit and then there's more bullshit. I think there are a lot fewer chemovars out there, a lot fewer landraces were used to create, primarily, what we've got."

"Like Northern Lights."

He folded his arms across his chest and laughed.

"Northern Lights is like that ugly cat that made all the cats in your neighborhood. I mean, there's a lot of Northern Lights in strains that people will never admit in a million years that it's there. You hear about OG Kush, you know, the Chemdog story . . ."

It seems to me like almost every strain has some kind of origination myth behind it, and the popular Chemdog strain is no different. As legend has it, the weed was first smoked by a Deadhead at a Grateful Dead show in the midwest. Someone named "Joebrand" sold an ounce of super-good weed to a guy who went by the name "Chemdog." Chemdog liked the weed and arranged for two more ounces to be shipped to him on the East Coast. One ounce was seedless but the other turned out to have thirteen seeds. I'm not a mystic, but there is apparently some kind of hippie voodoo significance to the Grateful Dead, the number thirteen, and these magic seeds.

Michael was skeptical.

"Now, maybe they found it at a Grateful Dead concert. Maybe. But you know, the guys who invented OG Kush invented it in Sylmar and they were breeding Mexican dirt weed and Northern Lights. That's where OG Kush came from. What's interesting is the chemovars that show up in OG Kush are Northern Lights, Thai, because only northern Thais smell like oranges, and Nepalese, because only Nepalese smell like fuel. So whatever junkyard dogs are in these strains, you know, somebody hit the jackpot and got the right phenotype."

I wondered if Michael had a favorite strain. Was there some

kind of holy grail of dankness that he was searching for with all this science?

"The chemotype I'm searching for—which is kind of gone now—is the classic Spruce Sativa Kona, the Big Island strain that we're pretty sure is a cross between Highland Oaxaca and Thai genetics. It didn't smell piney, it smelled sprucey; it really smelled like a Christmas tree. And an insanely clear effect. Just perfect. You had the high-functioning buzz that took the edge off everything and a lingering tickle of joy. It's like a pre-psychedelic tickle. That's what I'm looking for, and I'll find it."

"Is that why you started the collective?"

"I started Cornerstone because I was searching for the best. I mean, literally, I used to have to drive all over town looking for great meds, and they were really hard to find. I'd end up going to one shop for one thing and another shop for yet another thing to get my medicine chest and after a while I said 'Screw this. I'm going to start my own.' I thought I knew what I was doing, but I didn't know anything. But what's cool is that after four years of doing this I've learned a lot, but I realize that I'm only about twenty-five percent there. This game is in its infancy. Even though the plant has been around for five thousand years, the scientific breeding of cannabis as opposed to the informed farmer breeding of cannabis is in its infancy. It's really going to get interesting and I want to ride that train."

Chapter Fifteen

El Toro

Toronto reminds me of Shanghai. Not that I've been to Shanghai. Not that the streets of Toronto are festooned with signs in Mandarin. It's just that there seems to be a lot of building going on in Toronto and whenever I've seen pictures of Shanghai, that pulsing metropolis on the Huangpu always seems to have hundreds of construction cranes perched atop sleek high-rise buildings. Toronto seems new and vibrant. It looks modern, too, a futuristic city rising up alongside a large body of water just like Shanghai. Of course, I could've said Dubai—it also has sparkling new architecture and construction cranes and a large body of water—but Toronto seemed more like Shanghai. Not that I've been to Dubai.

I was in Toronto to attend the *Treating Yourself* Medical Marijuana and Hemp Expo, which was being held at the Metro Toronto Convention Centre.

Treating Yourself is a monthly magazine devoted to alternative medicine, including medical marijuana, published by Canadian activist Marco Renda. Canada has a well-established underground cannabis culture, but with legal medical marijuana beginning to take off there, it was suddenly a new and lucrative market that the major seed companies from Amsterdam couldn't ignore.

Green House Seeds brought their A game, setting up an exhibition booth that was twice the size of any other booth on the trade show floor. The walls of the booth featured large flatscreen monitors playing endless loops of Arjan and Franco in their *Strain Hunters* costumes looking like rugged adventurers as they trudged through tropical landscapes. A large section was devoted to selling T-shirts and hoodies emblazoned with the Green House logo and other merch, and there was a VIP area that looked like a typical suburban living room.

Arjan and Franco hung around the booth, greeting people as they walked by, discussing their strains and offering advice to growers, while attractive young women—local actresses hired for the convention—handed out thousands of free plastic grinders, posters, and seed catalogs.

DNA Genetics was there as well. They had a booth about a quarter the size Green House's, but cozier, with a trophy case displaying awards from the various cups and competitions they've won. They also had racks hung with various DNA logo wear and a cushy couch and chair set up behind the counter so guests could sit and chat with Don and Aaron.

Aaron waved to me as I approached the booth. He was in the middle of a scrum of about a dozen Canadian growers engaging in a stoner version of show and tell. The growers all had the same look—like your average fan at a rock show—wearing a uniform of T-shirts, jeans, and tattoos. They crowded around Don and Aaron jostling for position, pulling their best buds out of plastic baggies and presenting them for inspection. The buds were sniffed and eyeballed, snapped and squeezed, as Don and Aaron offered critiques and growing advice. It was like watching expert gemologists grade diamonds and rubies in a crowded airport terminal. One bag was "well grown," while another was "really well grown." Every now and again Don would sniff a bud and smile before tucking it away in his pocket. Those, I can only assume, were the super well grown.

As this mayhem was transpiring, the buds flashing fast and furious, Aaron's wife, Kim, sat serenely in one of the overstuffed chairs breast-feeding their baby.

Because I'd been talking with Aaron, a grower mistook me for someone who knew something about growing weed and asked if I had any experience with the Chocolope strain. I joked that they were much smaller than Jackalopes—the mythical half rabbit, half antelope of the West Texas wastelands—and a hell of a lot harder to catch. He gave me a blank stare and said, "Somebody's smoked too much."

Aaron laughed. "And then you gotta milk 'em, bro. It's not easy to milk a Chocolope."

A grower from Quebec, a lean and muscular man whose name reminded me of a French vintner, pulled out a Tupperware container packed with hash he'd sifted himself. Aaron picked a clump up and rubbed it between his fingers.

"You should use a finer screen, bro. Get some of the debris out."

The grower shrugged. "I don't really like hash. So I don't care."

Aaron recoiled. He looked personally offended. "If you're not going to care, don't bother, bro. Don't waste your time."

It's strange attending a convention for an underground and, in most cases, illegal industry. But I've been to a half dozen of these things. Large-scale trade shows like THC Expose in Los Angeles or Spannabis in Barcelona draw upward of thirty thousand attendees; other events are funkier, like the sparsely attended HempCon in L.A. where tattoo artists and T-shirt vendors outnumbered the cannabis companies.

It's not just seed companies that exhibit at these conventions: Glass blowers and bong makers—including the world-famous RooR bong company from Germany—attend, too, as well as vaporizer manufacturers, rolling paper manufacturers, hemp clothing companies HoodLamb and Ha-Swësh and others, political action groups, medical marijuana dispensaries, and companies that manufacture grow lights, watering systems, grow boxes, hydroponic equipment, soils, and assorted fertilizers and nutrients. The conventions traditionally host speakers and panels on a variety of topics ranging from horticultural tips and cooking with cannabis to political activism and recent medical breakthroughs. They are a strange cross between farming expos, fashion shows, and counterculture conclaves.

The *Treating Yourself* Expo was on the small side, like the first *High Times* Medical Marijuana Cup in San Francisco, but I like these smaller expos; they're low-key and friendly. The convention in Toronto also featured a film series, with documentaries celebrating the medicinal properties of cannabis, a biography of the late hemp activist Jack Herer, and the world premiere of

Strain Hunters: India Expedition, featuring a Q&A session with Arjan and Franco after the screening.

Smoking reefer is forbidden at most of these conventions. Typically, there is no herb on display or seeds for sale, and the only way you can get high is to sneak outside and blaze at your own risk. But there are a few exceptions. The *High Times* Medical Marijuana Cup in San Francisco had a smoking patio where patients with valid doctor's recommendations could smoke. If you didn't have a doctor's recommendation, they had a doctor standing by who could assess you. Somehow the event in Toronto had managed to get permission to build "the world's largest vapor lounge" in the Metro Convention Centre. It was large: four thousand square feet of vaporizer central. There were a dozen Volcano vaporizers manned by volunteers who were eager to help anyone with a medical marijuana card medicate safely on the premises.

Aaron had given me a bud of Chocolope to take to the vapor lounge to sample. After my medical certificate was checked, my California recommendation passing muster in Canada, I was handed a personal vape bag and mouthpiece, a package that was neatly self-contained in a clear plastic pouch with a cord that let you hang it around your neck. A gracious vaporista cheerfully ground up the bud and cooked me a big bag of Chocolope mist.

Like Sleestak, Chocolope is one of my favorite DNA strains, and the quality of this bud, grown indoors by a Toronto grower, was excellent. The vaporizer allowed the flavors of the strain to come through, and I got a nice taste of the earthy cocoa-flavor from the Chocolate Thai parent.

I exhaled and looked around the vapor lounge. It wasn't crowded—maybe twenty people were standing around talking and taking occasional hits off their vapor bags. The conversation was about strains and various flavors and effects. It was a convivial group of connoisseurs.

Ironically, this is the same convention center where less than a month earlier the G20, a group of first world countries united by economic and security interests, held its summit meeting while antiglobalism activists and anarchists rioted in the streets outside. I took another pull on the bag and suddenly wished I had access to a time machine. I'd love to send the entire vapor lounge back in time. What would the G20 summiteers make of this weed tasting? Would it be cool to suck down a vape bag of Super Lemon Haze with the Chinese premier Hu Jintao and French president Nicolas Sarkozy? Would German chancellor Angela Merkel think the Chocolope was dank? I imagine President Obama wouldn't have minded too much, since he famously admitted to smoking weed while he was in college. Maybe we could cook up a bag of Sour Diesel and talk about economic justice with Cristina Kirchner of Argentina.

But I think that if I had to choose one world leader to get high with, it would have to be Silvio Berlusconi of Italy. He is apparently no stranger to weed, wine, and teenage hookers. I may not like his politics but, let's face it, the dude knows how to party.

I finished the vape bag and realized I'd had maybe a little too much Chocolope, so I left the lounge only to discover that the universe has a wicked sense of humor. As I emerged from the haze of the vapor lounge, I ran into a large group of people waiting to audition for a reality TV show called *Wipeout Canada*. The premise of the show is to send contestants through a ridiculously difficult obstacle course designed to make them wipe out in the wettest, slimiest, most humiliating scrotum-smashing way possible for the amusement of the television viewing audience.

The *Wipeout Canada* aspirants were dressed up in a variety of costumes. I saw a woman dressed like a robot, a guy in a bee suit, someone wearing a dinosaur outfit, various spandex abominations designed to highlight the fitness, or lack thereof, of the wearers, and a long-haired dude wearing a checkered tablecloth as a cape.

Not far from this freak show was the lobby of the Intercontinental Hotel, which was hosting the pit crews and racing teams of the Honda IndyCar race that was happening that weekend.

I walked past the *Wipeout* wannabes into a crowded lobby full of earnest young gearheads wearing polo shirts designed to look like the logo-clotted racing suits of their respective teams. The gearheads didn't mingle. They were divided into logo tribes, matching teams moving through the lobby in tight huddles, talking in solemn voices about time trials, tire treads, and the weather.

Between the costumed freaks and the chassis chatter, I had to make an immediate exit.

I hit the streets, heading out into the warm evening air.

Lake Ontario stretched out to the horizon, reflecting the sunset, as daylight faded and the electric lights of downtown Toronto blinked to life. Toronto is a surprisingly beautiful city.

I was walking up Blue Jays Way, past the sizzle and stench of the hot dog vendors, when I suddenly realized that something was wrong. Very wrong. I stopped and looked at the people on the street and realized that almost every one of them was wearing a Rush T-shirt. It was like that moment in a zombie movie when the hero recognizes that things have changed. Everywhere I looked there were clusters of dudes with mullets wearing shirts that said "Fly by Night" or "Rush 2112." I had heard about the Canadian content laws that require a certain percentage of all TV and radio broadcasts to involve Canadian musicians and writers, but this seemed absurd. I tried to rationalize it, thinking that maybe the residents of Toronto were just really into Rush. I mean, somebody must like that music.

But then the number of people in Rush T-shirts began to grow exponentially, like some kind of malevolent magic trick, dividing and growing like the mops gone out of control in *The Sorcerer's Apprentice*. There were now thousands of them, bunching together like herd animals, an army of pale and acne-scarred teenagers, crossing the street toward me.

I was on the verge of a panic attack. What did these people want? Where were they going?

An affable black man walked up to me. He was not wearing a Rush T-shirt.

"You need tickets?"

"To what?" I asked.

He nodded, indicating the building behind me. I turned, followed his look, and, in that moment, realized that Rush was playing a concert that night and I was standing in front of the venue.

The hotel concierge had told me to walk up to King Street where I would find "all kinds of restaurants." I walked up toward King Street, passing a number of nightclubs and bars. That's not really surprising for a downtown street in a major metropolis, but I was baffled by the complete lack of imagination on display in the names of the clubs. There was Loft, Light, Reign, Marquis, Tryst, and Time. But the concierge was right about King Street: It was restaurant row. Completely by accident I ended up at a place called Lee. The menu claimed that they served "Chinese tapas." I like tapas and I like Chinese food, but when I think of the two together I usually call it "dim sum." But the menu looked interesting—not a *shu, mai,* or *bao* on it— and the restaurant was packed; always a good sign.

The hostess had room for one at the bar, and I didn't hesitate. I didn't know it at the time, but it turned out that Lee is run by Top Chef Master Susur Lee and is one of the most popular restaurants in Toronto. Getting a seat on a Friday night was, well, really lucky.

I was still in the midst of a Chocolope high, which made reading the menu problematic. It had nothing to do with my eyes; *everything looked delicious.* I decided to leave it up to the waiter ("Dazzle me!") and sat back with a cold glass of white wine to take in the scene.

It was after the second bite of the crispy tofu rolls glazed with a honey-soya-chili sauce that the situational aspects really hit home. The Chocolope was dank, the crispy tofu rolls were dank, the music was groovy, the waiters friendly, the white wine was cold and delicious, and all of it together elevated the experience. It could be argued that the cannabis had heightened my senses—that my organoleptic experience of tasting the coconut-lime soup or listening to the music in the restaurant was enhanced. I would agree with that, but at the same time I think that the quality of the food—the complex flavor profiles combined with near perfect execution in cooking—worked in a similar way. If the Chocolope enhanced the experience of the food, the food enhanced the experience of the Chocolope.

I turned and watched as a plate of slow-braised beef was delivered to the couple at the bar next to me. The woman took a bite of it and groaned with pleasure. She looked at her husband and said, "Oh my God. I just had an orgasm. A real orgasm."

I wondered what would've happened if I'd gone to the Rush concert.

The next afternoon Don and Aaron introduced me to a young Canadian who looked like an elven ranger from the World of Warcraft computer game. He was small and compensated for his lack of size by cloaking himself in a black leather jacket with chrome chains dangling from the epaulets. He smiled a lot, but his seemingly friendly demeanor was undercut by intense dark eyes and a black pinstripe beard that gave his face a demonic appearance.

Aaron waved me over.

"Check it out, bro."

The elven ranger smiled and carefully opened a small jar, as if it were the world's best caviar, and showed me a glistening pool of viscous oil. Oil—or hash oil, as it's sometimes called—is pure, concentrated resin with a THC content of 70 to 90 percent. It is super-potent stuff. Traditional oil smokers put a drop on a metal disc and then inhale the smoke that comes off the burning oil as the disc is heated. Nowadays there are complicated glass contraptions that look like water pipes or "percolator" bongs with built-in filters, but they still use the heated disk to cook the oil. I was invited to come along and try it, but I declined. It was early in the day and the oil looked malevolent, like some kind of elven alchemy.

Instead I went and sat in the VIP area at the Green House booth and watched Franco and Arjan demolish heaps of Chinese takeout in near silence. Franco expertly wielded his chopsticks, quickly shoveling heaps of vegetables and fried rice into his mouth. He stopped for a moment, chewed thoughtfully, and offered a kind of apology.

"I'm a skinny motherfucker and get low blood sugar."

One of the local models hired to hand out freebies came up to them and looked at the boxes of takeout. Arjan smiled at her. "Help yourself."

She lifted the lid on one of the Styrofoam containers and made a face like she'd just smelled a fart.

"I'm a vegetarian."

She dropped the lid and walked off.

The VIP area in the Green House booth wasn't private. It was really just a couch, coffee table, and a couple of chairs arranged like a living room, out in the open, with a little velvet rope separating it from the rest of the convention. It resembled a set for a domestic sitcom with an audience of gawkers standing on the perimeter and yet, once you were in it, it felt oddly private.

The filmmaker Melissa Balin joined us. She's a bright and energetic woman with untamed hair that flops off her head and into her face when she talks. She is, like all filmmakers, a natural hustler. Melissa was attempting to organize some scenes for her documentary and wanted to get footage of Arjan and Franco on something called the "Canna Cruise." The convention organizers had rented a party boat to putter around on Lake Ontario and called the event "Hemping on the High Seas." It was a VIP shindig that promised an "extravagant dinner and unlimited nonalcoholic drinks."

Melissa knelt on the floor by the coffee table, scrupulously avoiding the Styrofoam containers of food as she launched into her pitch. "I was thinking about getting you guys on the boat, maybe you could be involved with the judging, something like that."

Arjan, his mouth full of noodles, wagged a finger at her. "No. Absolutely not. No pictures of us smoking."

Melissa nodded and brushed some wayward curls behind her ear as she recalibrated her pitch with this new information. "Maybe just you guys and a couple of vaporizers in the background."

Arjan took a swig of beer and cleared his throat. "We're not going to be smoking, next to people smoking, nothing like that."

I don't think she was expecting this but, to her credit, she didn't let it throw her. "Okay. Well then maybe I could get a segment with Franco talking about Bullrider."

Bullrider is a strain that Franco really likes, but before Franco could say anything Arjan shook his head. "No."

I could see that Melissa was getting frustrated, but she smiled good-naturedly. "Okay, well. We'll figure something out."

Arjan nodded and she stood and walked out of the VIP area. He slumped back on the couch and watched her go with a baffled look on his face. He turned to me and Franco. "She thinks we're gonna go on film saying we smoked? We're foreigners in Canada. We don't want pictures of us doing anything illegal. They'll never let us back."

Although it annoyed Melissa, I understood Arjan's concerns. He's a savvy businessman and understands the importance of portraying himself and his company as a responsible, some would say corporate, enterprise.

I asked if he enjoyed coming to these trade shows. He nodded. "Of course. But I think we did too many last year. Next year I want to do only three."

Franco nodded as well. "The big ones, for sure. Barcelona and the Cannabis Cup."

"Hey," I said, "speaking of the Cannabis Cup . . . what're you guys entering this year?"

In the past, whenever I'd pressed Franco for a hint as to what they might enter, he would talk about growing and selecting the best herb and the importance of perfectly curing it before making a final decision. But when I asked Arjan, he confirmed what I suspected.

"It's been decided. We'll try to make Super Lemon Haze the three-time champion."

When I returned to the DNA Genetics booth I found Kim holding the baby as she fielded complicated questions from various growers about the care and feeding of DNA's popular strains. She obviously knew what she was talking about—she's as much an expert as her husband or Don. As she chatted to the grower, I heard a soft groan coming from behind her. I looked over and saw Aaron splayed out on the couch like an old coat.

The color had drained out of his face and he looked shaky and exhausted.

I asked him if he felt okay. He blinked up at me.

"That fucking oil, bro."

Aaron pulled his baseball cap down over his eyes and his head lolled to the side. He rolled with it, curling up in a fetal position. His voice turned to a faint falsetto as he turned away and attempted to burrow into the couch. "I'm so high right now. I am never going to smoke weed again."

I made my way back down to the vapor lounge to try a strain called Lavender that a grower from Vancouver had presented to Franco. Franco was kind enough to give me a huge bud of it, informing me that this was from one of the best growers in all of British Columbia.

The bud was an unusual color, almost pale violet, and so sparkly with trichomes that it looked like a hunk of amethyst. I entered the vapor lounge and took my vape bag out of its plastic case. I found a friendly vaporista and pulled out the Lavender. The size of the nugget alone might've caused a stir—it was large, dense, and perfectly formed—but the color elicited gasps of excitement from the people standing nearby.

Lavender is a strain developed by Soma of Soma Seeds in Amsterdam and is a cross of Super Skunk, Big Skunk Korean, and a Hawaiian-Afghani blend. It was voted Best Indica at the 2005 Cannabis Cup. Of course I'm not an indica smoker; I don't react well to it. Franco once speculated that I have some sort of metabolic problem with indicas. But then, at the time, I didn't know that Lavender was predominantly indica and Franco must've forgotten about my metabolic reaction when he gave it to me. But I was excited to try one of Soma's strains.

In the world of underground botanists Soma is a bit of an anomaly. He's an iconoclastic Caucasian Rastafarian in his late fifties with large sprays of gray dreadlocks that cascade down his head and mix with his shaggy gray beard. He looks a lot like a friendly English sheepdog.

Soma doesn't raise cannabis for cash and prizes. He's not in it for brand identity or market share—in fact, he believes that a grower's intention is just as important as lighting and fertilizer because "money vibes" can harm cannabis plants. Soma is on a spiritual mission, and he walks it like he talks it. He's a vegetarian, wears only clothing made out of hemp, smokes daily, and truly believes that cannabis can save mankind from destruction. I don't know if cannabis can save mankind, but it sure can't hurt.

He's also the botanist that developed NYC Diesel, which can, occasionally, smell like bus exhaust on Canal Street. NYC Diesel is a cross between Mexican sativa and Afghani and has placed second in the Best Sativa category at three different Cannabis Cups. He is obviously a strain breeder who knows what he's doing.

A crowd gathered as the Lavender was vaporized, and I broke off chunks of it and passed it around. I gave a chunk to a very stoned middle-aged woman who winked at me and then leaned in and whispered, "Do you want to see my skunk?"

I didn't know how to respond. *Does a gentleman look at a lady's skunk?*

But before she could flash her skunk, I was handed my vape bag filled with Lavender mist. I was immediately impressed by the flavor. It was spicy and subtle and actually tasted a little bit like some lavender-infused ice cream I once tried. Other people in the lounge began vaporizing their chunks of the herb, and I have to say that Lavender was a runaway hit. Everyone loved it. They were raving about the potency, the taste, the effect. Unfortunately, I had my typical reaction to indica and became woozy and somewhat paranoid.

I spent the rest of the afternoon in my hotel room gazing out the window at the view of the convention center roof.

I wasn't surprised that the Lavender was so good. It was from, as Franco said, "one of the best growers in Vancouver," and the city of Vancouver—in fact, the entire province of British Columbia—is home to some of the best growers and strain developers in the world. The casual community of cannabis entrepreneurs in the region has formed what they call "the Union." There is even a documentary film by the same name that explores the pervasiveness of the culture and the positive influence cannabis cultivation has on the Canadian economy. In BC, standards are high and the competition is fierce. Because of this regional devotion to producing top-quality weed, BC Bud has become a kind of brand name, an underground *appellation d'origine contrôlée,* the term the French government grants to certain agricultural products to ensure their quality and origin. This kind of unofficial AOC also applies to regions such as California's Humboldt County or the Big Island of Hawaii. This is not to say that there aren't skilled growers in Toronto, Montreal, or any other Canadian town. The whole country appears to be in the midst of a cannabis renaissance.

A number of Canadian politicians, most notably Vancouver's former mayor Larry Campbell, have suggested that cannabis would be legal in Canada if it weren't for political pressure from the U.S. government. In a 2007 interview in the Vancouver *Province,* Campbell expanded on this idea. "It's all ideology," he said. "If they're wrong on this, then what else are they wrong on? They won't even allow hemp. That's how stupid these people are—and they are stupid. I describe [then White House drug czar John] Walters as a moron, and he is truly a moron."

The behavior of the U.S. government doesn't stop with mere bullying. In 2005, in an act of what can only be called overreach by the DEA and sycophantic brownnosing by Canada, the Royal Canadian Mounted Police arrested Marc Scott Emery, the self-proclaimed "Prince of Pot." His crime? He had mailed cannabis seeds to customers in the United States.

What's interesting about this is that while selling cannabis seeds is technically illegal in Canada, it is only punishable by a small fine. Yet the U.S. government convinced a foreign country to detain one of its own citizens and turn him over to face much harsher penalties in the United States. Why? I think the answer can be found in the original text of DEA administrator Karen Tandy's statement released on July 29, 2005:

Today's DEA arrest of Marc Scott Emery, publisher of *Cannabis Culture* magazine, and the founder of a marijuana legalization group—is a significant blow not only to the marijuana trafficking trade in the U.S. and Canada, but also to the marijuana legalization movement.

His marijuana trade and propagandist marijuana magazine have generated nearly $5 million a year in profits that bolstered his trafficking efforts, but those have gone up in smoke today.

Emery and his organization had been designated as one of the Attorney General's most wanted international drug trafficking organizational targets—one of only 46 in the world and the only one from Canada.

Hundreds of thousands of dollars of Emery's illicit profits are known to have been channeled to marijuana legalization groups active in the United States and Canada. Drug legalization lobbyists now have one less pot of money to rely on.

I didn't realize that political donations and the hiring of lobbyists to promote legislation was against the law. It can't be because he's a foreigner trying to influence U.S. policy; former assistant secretary of defense in the Reagan administration and Dick Cheney insider Richard Perle was on Libya's payroll as a lobbyist. I don't see him being extradited.

Clearly Marc Scott Emery is a political prisoner, a POW of the war on drugs. He was put on the "most wanted list" not because of the nature of his crime, but because he was helping to fund the opposition.

While he was out on bail, fighting extradition, he and I exchanged emails. I was planning on going to Vancouver to talk to him when suddenly the Canadian government signed his extradition papers and he was taken into custody and sent to the United States. He's now serving five years at the Yazoo City Federal Correctional Complex in Mississippi.

That evening, refreshed from my indica-infused nap and emboldened by a cold Canadian beer, I attended the *Treating Yourself* Medical Marijuana Cup Awards, presented in association with the Marijuana Music Awards.

The event was being held in the John Bassett Theatre in the

convention center, a large and modern theater with comfortable seats and air-conditioning that kept the room at sub-arctic temperatures. Despite seating for more than a thousand, there were only thirty or forty people attending the awards, and the sparse crowd gave me the feeling that I was at a dress rehearsal for an awards show and not the real thing.

A band was playing when I arrived and, for a minute or two, I couldn't tell if they were culled from the auditions for *Wipeout Canada*. There was a handsome guy dressed in jeans and a flannel shirt with a woolen beanie on his head singing a song about having sex with a truck driver, while a dominatrix—complete with riding crop—and Catholic schoolgirl sang backup. It was generic rock at its most generic, especially when they sang some mock rap song about "the chronic." I was both baffled and delighted by the lead guitarist, a man clad in black goth warrior armor with a gimp mask and what looked like a leather sombrero stuck around his neck.

The curtain dropped on the band and an elderly gentleman came out on stage dressed in jeans and a T-shirt, as if he'd just been pulled from the audience and handed a microphone. He was the awards emcee and began by announcing that he had forgotten his reading glasses. He held his script up a few inches from his nose and announced the winners. For unknown reasons, the stage manager had set the table full of trophies on the opposite side of the stage, so the myopic emcee would announce the award and then have to scurry to the other side of the stage, pick up the trophy, and race back to the podium to hand it to the winner. These sprints would leave him gasping for air, and he took several moments to catch his breath before he could begin reading again.

The Song of the Year Award went to Chief Greenbud for "It's 4:20 Somewhere," a parody of the hit song "It's 5 O'Clock Somewhere" made famous by Alan Jackson and Jimmy Buffett. I wondered how much of a parody it was because it didn't seem like much of a stretch to move the clock back forty minutes and call it a new song.

I should make it clear that the chief didn't appear to have any relation to the indigenous people of the Americas; he is, by all appearances, a pudgy middle-aged Caucasian from Nashville; but Chief Greenbud was humble and charming, graciously thanking his wife and kids for their support. He announced that he'd be back shortly to perform the song for us.

The first Medical Marijuana Award of the night was the Seed Company Sativa category, presented to Serious Seeds for their AK-47 strain. A tall Dutchman named Simon came up to accept the trophy, which was a very lovely glass bong designed by RooR. Simon blinked out at the audience, obviously surprised by his victory.

"I just sent some seeds to a grower here. I had no control over it." He shrugged and in a burst of compulsive honesty announced, "I didn't really like it when it came in so . . . thank you, judges!"

He held the bong up in the air and beamed.

There were more musical interludes. An attractive dreadlocked woman named Sahra Indio rapped sweetly over a reggae sound track, but the lyrics to her song "Spiritual Connection" sounded as if they were cribbed from a self-help manual cowritten by Tony Robbins, Oprah Winfrey, and the Dalai Lama. "Be still. In the silence, wisdom is revealed."

I suddenly, desperately wanted another cold Canadian beer. A few people sitting behind me got up and left. Perhaps they'd read my mind and were going to the bar.

I was pleased to see the affable Chief Greenbud return to the stage. He stood in front of a microphone, just a man and his acoustic guitar, and performed the Song of the Year for us. I will say that the chief has a deep, country baritone that sounded quite capable of churning out Nashville hits, but when he started to sing a second song about getting arrested and asking the cops for a "reach around," and as more people in the upper levels of the auditorium discreetly slunk out the door, I decided that my desire for a beer trumped any further journalistic responsibilities.

I'm a fan of the *Treating Yourself* Medical Marijuana and Hemp Expo. It had a kind of earnest community theater charm. It wasn't as hectic or well attended as the expos in Los Angeles or Amsterdam, but it was friendlier.

But for me, the real highlight of the convention was meeting the legendary strain developer named Reeferman. Reeferman is an underground botanist from Canada, and the man who created such famous strains as the Cannabis Cup–winning Love Potion #1, the first-ever Cup win by a Canadian, and Willie Nelson, a strain that was built to the wiry Texan's specifications. He's won a bunch of other awards and was inducted into the *High Times* Seed Bank Hall of Fame in 2007. On a personal note, he's the botanist who developed the John Sinclair strain—which turns out to be an equatorial sativa from Congo—so in a way, he's responsible for me even wondering what dank was in the first place.

Considering how long he'd been involved in the seed business, Reeferman was a lot younger than I thought he would be. He was also bigger. He's probably six foot three and on the hefty side. When he stood in the small booth, with his thick head of hair, full beard, and piercing blue eyes, he gave off a kind of *Ursus americanus* vibe, although once he started talking he was actually soft spoken and charming, shy almost, making him seem more like a teddy bear than a real bear.

He was also super articulate and passionate when it came to matters involving cannabis and cannabis breeding. Unlike self-taught breeders Don and Aaron, Reeferman has a bachelor of science degree in agriculture and worked as a researcher for the Canadian Department of Soil Science at the Agassiz research station in British Columbia's Fraser Valley. With his training, experience, and unapologetic love for cannabis ("I just love the way the plant smells," he said. "I like the smell of it better than smoking it."), it was only a matter of time before he turned his attention to full-time strain breeding and seed production.

I was surprised to see him at the convention. I had heard rumors that he'd quit the scene and stopped producing seeds. He was supposedly in a self-imposed exile, his strains unavailable. The truth was much worse than the rumors. A couple of years ago, the Canadian police raided his Saskatchewan farm, confiscated his land and livestock, left his wife and kids homeless, and threw him in jail for illegal cannabis cultivation. He served four months of his sentence in the oldest, and worst, prison in Canada before being released.

When he got out, Reeferman was left with nothing. He left Canada and started over in Mexico, finding the Vicente Fox administration tolerant of cannabis agriculture and commerce. When Felipe Calderón became president, his government expressed a less supportive view.

Reeferman moved again, this time to Colombia to set up greenhouses. But South America isn't a great climate for bear-like Canadians, so when the Canadian government began issuing licenses for legal medical marijuana growers, Reeferman saw it as a call to return to his homeland. He now has one hundred acres of agricultural land in a small town called Moose Jaw.

Reeferman speaks with a soft, assured, and thickly Canadian voice. He admitted he doesn't attend many conventions and wasn't planning to come to this one, but it was "the first one in Canada and I wanted to be supportive." He also had good business reasons. He was introducing a new line of high-end liquid nutrients called "Love Potion," produced by his company, Maple Reef Plant Products, and he was at the expo with his new sales rep, a British seed distributor called Seedsman.

I asked him if he was planning on entering the Cannabis Cup. He nodded. "Yeah. I've got a strain called Cherry Haze that I'm thinking about entering."

There it is. A fruit flavor mixed with Haze. Coming from a breeder as talented as Reeferman, BC Bud has a good chance of causing an upset in Amsterdam.

Chapter Sixteen

Sticky Fingers

"Dude, do you know the pterodactyl?"

Red rocked back in his chair, a large bud of freshly dried Island Sweet Skunk in one hand, a pair of small Fiskars scissors in the other, and a big goofy grin spreading across his face.

"You mean the flying dinosaur?" I asked.

Red's face flushed. He shook his head and laughed.

"No. No, it's a . . ." He stopped short, his face glowing even brighter. "Guess."

Red giggled and blushed some more, rocked forward on his chair, and added the bud to a pile of freshly trimmed buds that lay on the table in front of him. Based on the fact that his face had turned tomato red, I made an educated guess. "Is it some kind of sex thing?"

Red stammered. "It's a . . . you . . . ha . . ."

He couldn't bring himself to tell me what kind of pterodactyl he was talking about.

Luckily, one of the other trimmers in the room was happy to oblige. A portly young Latino named Chuva leaped out of his chair. "It's like this, man."

He stood and spread his arms, like he was being frisked by the police.

"You got one guy in front, suckin' you off, while another guy comes up and does you from behind." Chuva waved his arms, adding a flourish, then stood perfectly still like a gymnast who's just stuck a landing, as if it's the simplest thing in the world. "The pterodactyl."

Chuva took a bow as Red erupted into a fit of laughter, cackling and rocking back dangerously, his legs pumping up and down in delight. Chuva smiled, pleased with his performance, and then plopped back in his chair, picked up a bud, and resumed trimming.

So this was a trimming operation.

It had been my intention to come back to the Sierra Nevada to help Crockett with his outdoor harvest, but the threat of thunderstorms on Monday had meant that he had been forced to pull all but a few plants early. That's the gamble that outdoor farmers face: The plants were ripe, ready to be harvested, and yet another sunny day would've made them even better. Now, on Wednesday, the majority of the plants were hanging upside down on wires in a large, open barn, drying before being trimmed.

Crockett, who'd been busy weighing the trimmed buds and sorting them into cardboard bins while the pterodactyl demonstration was occurring, looked at me. "You see why I don't spend a lot of time in this room."

It wasn't really a room. The trimming operation was set up in what might've been an old garage but was now a fine example of the architectural style know as rural slapdash.

In addition to Red and Chuva, the trim team included Quay and R.J. They would be joined in a day or two by Cletus "The Dingo" McKlusky and someone named Shlev Dog. They were all young men in their twenties, hipster migrant workers, gypsy stoners, members of a loose fraternity of experts in the art of taking twelve-foot-tall cannabis plants and cutting away everything but the gleaming crystal-coated buds. To say that there's a demand for skilled trimmers in California is an understatement. Between outdoor crops harvested in the fall and indoor grow rooms constantly cycling new plants, expert bud barbers can be as busy as they want to be.

The plants hung behind them, upside down, like sides of beef, and the trimmers took turns breaking them down—stripping away the stalks and stems. They separated the buds from the stems and then meticulously manicured them, snipping off as much of the leaf as they could before putting the trimmed buds into special drying racks. Any part of the plant that wasn't the dried buds was swept off the table and into a large clean cooler. It's important to keep these excess leaves clean and dust free until they can be processed into hash. Nothing is wasted. The room was, for all intents and purposes, a cannabis abattoir.

Individual work lights illuminated each man at his station, as they kept one eye on the task at hand and another on the flicker from a TV set that seemed to constantly play episodes of *Beavis and Butt-head.*

Chuva tried to teach me how to trim. "Look. Keep the scissors in one spot and move the bud."

He's an expert. His scissors moved in rapid-fire snips, like a sewing machine. He looked up at me without breaking from his rhythm. "I don't even need to look. It's all feel."

I sat down and tried it. The buds had more resin than I expected, gumming up the scissors and sticking to my fingers. It was like trying to eat pancakes and syrup with your hands. My first attempt at trimming looked distinctly like a squashed tater tot. I held it up for Chuva.

"How's this?"

Chuva shook his head. "You're too careful. You got to get in there."

He demonstrated, deftly buzzing off the leaves and reshaping my pathetic attempt into a perfect crystal-coated bud. I picked up another stem and tried again.

A good trimmer can make two to four hundred dollars a day, depending on how much they trim. The work is paid by the pound, and an expert can churn out two pounds of bud in a ten- to twelve-hour day. There are trim crews that travel around, working together job to job—I've heard of a famous all-lesbian trim crew that does exquisite work—and then there are the ronin—masterless samurai—trimmers who wander the countryside with scissors in their pockets, ready to pick up whatever work comes their way.

A harvest like the one Crockett had brought in, eighty to ninety plants, would take a team of trimmers a couple of weeks to process. It's monotonous work, akin to gutting fish at a cannery, but the boredom is relieved by an unlimited supply of premium cannabis to smoke, by *Beavis and Butt-Head,* and by jokes about imaginary sexual positions.

The trimmers were all staying in the farmhouse and Crockett had hired a cook to keep them well fed and happy. There was even a snack table filled with cookies, crackers, gum, sodas, and beer, similar to a craft service table you'd find on a movie set.

"I do whatever I can to make their life easy," Crockett said.

That's because no matter how good the genetics are, no matter how much care was taken to grow the plants, if they're trimmed to look like lumpy and deformed tater tots—if they don't have "bag appeal"—then Crockett isn't going to get a good price for his crop.

Farmers who grow outdoors have to battle gophers, bears, the weather, mold, insects, DEA agents in helicopters, and just bad luck. To survive all that, harvest a bunch of healthy plants, and then have an amateur trimmer destroy the buds is unthinkable.

This was what was on my mind as I mangled another bud. I looked over at Crockett. He was pretending not to notice, but I could tell that he had a calculator running inside his head, watching his profit margin diminish with every snip of my scissors.

Chuva decided to take a break and loaded some dry-sift hash into a bong. He had scraped his scissors every now and then, collecting blobs of resin he called "scissor hash." He looked at me. "You smoke hash?"

I nodded.

"You smoke weed?"

"Of course."

Chuva shook his head. I had somehow disappointed him.

"I don't smoke weed," he said.

He flicked his lighter and ignited the hash, taking in a huge, if somewhat contradictory, hit.

"Why not?"

Chuva exhaled and jumped up from his chair shouting, "Because plant material fucks you up!"

He put the bong down, sat back in his chair, and fixed me with a glassy look. "You know what I'm saying?"

"Not really."

Chuva settled into his seat and adopted a professorial tone. "Take tobacco. It's plant material." He spread his hands out, palms uplifted in a questioning shrug. "What does it do?" he asked.

"Give you cancer?" I said.

He nodded. "It irritates your throat, irritates your lungs. Plant material fucks you up."

Chuva looked over at the other trimmers and cocked an eyebrow. "What is cancer?"

He was asking the entire room. I know a rhetorical question when I hear one so instead of saying something about malignant cells, I just watched. Chuva waited, making sure we were all paying attention, and then continued.

"I'll tell you what it is. Cancer is an irritation that took on a life of its own. And that's why I don't smoke weed." He picked up a bud and resumed trimming. "Plant material fucks you up."

The sound of a car crunching on the gravel driveway interrupted us, and Crockett waved for me to come with him. We were both happy for me to stop destroying perfectly good buds, so I followed him out and watched as Slim and his wife, Natasha, pulled up in their SUV.

Slim uncoiled his lanky body, emerging from the car in stages. I'd forgotten how tall he is. Natasha said hello, then went into the house cradling a Crock-Pot. Slim and Crockett immediately set about rolling a joint. Crockett was worried that they needed more trimmers and suggested they bring down some people they knew from Humboldt County. Slim nodded thoughtfully; he was going to defer to Crockett on this.

Red's dog, a coyote-mutt hybrid, trotted toward us emitting a strange kazoo-like sound. At first I thought he was moaning or maybe he had a bone stuck in his throat, but then I realized that the strange sound was coming from the dog's ass. Slim laughed and turned to me, explaining. "He snuck into the house last night and ate two big pepperoni pizzas right off the table. I don't think he's feeling so good right now."

As if to offer proof of his misery, the dog lifted his tail into the shape of a question mark and began to fart some more. The small pops and sputters bursting from the dog's anus picked up momentum and the uninterrupted stream of canine flatulence lasted for almost thirty seconds. It sounded like a tire with a slow leak. The dog hung his head and whimpered.

Slim chuckled and shook his finger. "Maybe that'll teach you."

Crockett passed the joint to Slim and looked up at the sky. There was an ominous black cloud looming over the top of the mountains, and you could see gray sheets of cold rain draped along the ridge line. Earlier one of the park rangers had told me there was a small chance of snow or sleet—something that was not unusual for mid-October in the Sierras—and yet, oddly, it was warm and sunny where we were standing.

Crockett scratched his beard. Was he willing to let it ride? Or was he going to yank the last few plants? With each plant capable of producing several pounds of high-quality bud, there were thousands of dollars at risk by leaving them out in a potential storm.

Crockett turned to Slim.

"Jerry thinks the plants could stay in another couple days, but I think with this weather, we might want to pull 'em."

Slim shrugged. "Let's go take a look."

We walked down a dirt road toward a large stand of manzanitas that shielded the plants from high winds and snooping DEA helicopters. And, as if I were in some kind of recurring nightmare, Crockett told me to watch out for rattlesnakes as we moved off the trail and into the woods. I nodded. "Yeah, I heard it's a bad year for snakes." Slim offered this encouraging thought, "You never see 'em until you see 'em. I don't know why that is."

I suddenly missed Red and his shotgun.

We cut through the manzanitas and bear clover and reached a couple of large Island Sweet Skunk plants standing in the sun. The plants were easily ten feet tall, probably taller, with large *colas*—the Spanish word for "tail" that's used to describe the long clusters of buds—jutting out in every direction. The plants were majestic, really, healthy and green. They had attained their peak expression.

I was impressed by the plants, but what really struck me, something I hadn't experienced in an indoor grow room, was the smell. The scent coming off the buds was rich and strong and delicious. It reminded me of a musky, ripe pineapple.

Crockett and Slim examined the buds, checking the resin and looking for signs of mold. They decided they'd wait and see how the weather progressed. If it got bad, these were coming out right away.

I followed them through some pines and poison oak—the leaves now yellow—to a couple of beautiful Super Lemon Haze plants. The leaves were a vibrant green and the plants looked like Christmas trees, each with a large dominant *cola* rising off the top. These also gave off an intense aroma, but the Super Lemon Haze scent was distinct, spiked with robust citrus overtones. The plants don't smell sweet like roses or orange blossoms; cannabis fragrance has a different, deeper dimension. I finally understood why botanists Reeferman and Aaron of DNA say they like the smell of the plants just as much as the smoke.

Crockett squeezed the buds, checking for density. "These are close."

Slim came over and examined the plant. "But it could use another day."

I asked them if their clients, the medical marijuana dispensaries in L.A., paid more for the Super Lemon Haze. Crockett laughed.

"These aren't for sale. These are personal use only."

Crockett is part of a continuum that stretches back in time for almost ten thousand years. According to author Martin Booth in his exhaustively researched book *Cannabis: A History,* "Quite possibly, the first cannabis farming occurred in far western China or Chinese Turkestan. The oldest existing historical records and subsequent archaeological evidence, though not extensive, seem to support the theory that cannabis was being grown at the dawn of Chinese civilization."

With the stalks used for fiber to make paper, fabric, and rope; the seeds used for oil and food; and the flowers and leaves used for medicinal and spiritual practice, it's really no big surprise that this useful and versatile plant was cultivated by almost every culture in the world.

It helps that cannabis is incredibly hardy and grows almost anywhere. It can thrive in tropical and temperate climates, in poor soil, and with minimal water. It grows in equatorial jungles and at altitudes as high as eight thousand feet.

Cannabis is grown in the Himalayan highlands of India and the jungles of Cambodia; it's cultivated in the arid mountains of Morocco and the lush forests of Malawi; in Canada, Jamaica, and throughout Mexico, and in Central and South America. It's adapted to life on six out of seven continents, and where it's too frigid or there's not enough sun for an outdoor crop, the farmers move indoors and the plant adapts. Everywhere it grows, the U.S. government's "war on drugs" follows, turning farmers into outlaws and making the germination of a seed a criminal act.

It wasn't always this way. At the time of the American Revolution, cannabis—or hemp as it was called back then—was one of the principal crops grown in the colonies. The fiber was used for rope, cloth, and paper. According to hemp historian Jack Herer in his book *The Emperor Wears No Clothes,* "Until 1883, from 75–90% of all paper in the world was made with cannabis hemp fiber including that for books, Bibles, maps, paper money, stocks and bonds, etc."

As cannabis activists are quick to point out—and never tire of saying—the original Declaration of Independence was drafted on hemp paper. Hemp seed oil was commonly used in lamps and was pressed into a nutritious vegetable oil. As medical science advanced, *Cannabis indica* became one of the most common ingredients used in pharmacology to treat a wide range of symptoms for both humans and domesticated animals.

So what happened that turned the U.S. government against such a practical and useful plant? *People began using it for fun.*

Isn't that always the way? Once a party kicks off, some Puritan screams "Turn it down" and calls the cops. To make matters worse, it was "foreigners"—Mexicans and blacks—who used the plant for recreation.

In the mid-1930s, anti-immigration groups sprung up around the American Southwest. Looking for any excuse to slam the door on immigration from Mexico, they began spreading rumors about sex-crazed brown-skinned maniacs running amok on "loco weed."

Typical of this kind of fear mongering was a letter published in the *New York Times* on September 15, 1935, from C. M. Goethe, a member of a group called the "American Coalition" whose stated goal was to "Keep America American."

"Marihuana, perhaps now the most insidious of our narcotics, is a direct by-product of unrestricted Mexican immigration." The angry Sacramento resident then called for a quota on Mexican immigrants, stating, "Our nation has more than enough laborers."

Newspaper tycoon William Randolph Hearst, a man with a keen nose for the sensational, picked up the beat and began running articles warning the populace of the new drug menace perpetrated by, as Harry Anslinger, commissioner of the Federal Bureau of Narcotics, called them, "our degenerate Spanish-speaking residents."

Anslinger refined his Chicken Little act over time, amplifying the fear-mongering of Hearst and other "Keep America American" types in his testimony to Congress in 1935: "The deleterious, even vicious, qualities of the drug render it highly dangerous to the mind and body upon which it operates to destroy the will, cause one to lose the power of connected thought, producing imaginary delectable situations and gradually weakening the physical powers. Its use frequently leads to insanity."

You'd think he was talking about some kind of heroin/meth/LSD cocktail.

I don't know what was in the air on September 15, 1935—maybe a radioactive sunspot erupted and sent a burst of white-hot microwaves to fry people's brains, or perhaps it was simply Satan's birthday—but it's interesting to note that the same day the *New York Times* was printing letters from American crackpots denouncing Mexicans, in Nuremberg, Germany, at a Nazi Party rally, the Hitler government announced the enactment of the Nuremberg Race Laws. One of these was called "the Law for the Protection of German Blood and German Honor"—sounding like a formal order for frauleins to strap on chastity belts—that outlawed marriage and sexual intercourse between Jews and Germans. The Nazis cited "the provocative behavior of Jews" as the motivation.

Jack Herer and other cannabis historians have found evidence of collusion between large chemical companies such as DuPont and lobbyists for cotton growers that helped push anti-cannabis legislation forward. Eventually, the Marihuana Tax Act of 1937 was passed. The law required farmers to have a tax stamp to grow cannabis and, with the exception of a few rare cases during World War II, the stamps were never issued. Much like what happened when the Volstead Act outlawed alcohol—enriching Al Capone and other bootleggers—the war on drugs allowed vicious drug cartels to get filthy rich.

The war on cannabis may have its roots in racism, but when pot smoking moved from the jazz clubs of New Orleans to the beatnik coffee shops of Greenwich Village and college campuses across the country, the government used the law as a form of social control. The beat generation, hippies, and Black Panthers threatened the established social order, and drug laws were used as a cudgel to keep rebellious youth and dangerous radicals on the run and under arrest. Former president Richard M. Nixon echoed the political hysteria of the time in a conversation with his aides H. R. Haldeman and John Ehrlichman in the Oval Office on May 13, 1971: "You see, homosexuality, dope, immorality in general. These are the enemies of strong societies. That's why the Communists and the left-wingers are pushing the stuff. They're trying to destroy us." (Of course, Haldeman and Ehrlichman knew a lot about immorality. They later were indicted and sent to prison for their involvement in the Watergate scandal.)

But Nixon wasn't the last president to fear the cannabis plant. Ronald Reagan started the "Just Say No" campaign in the early 1980s, sending his wife around to schools to encourage students to abstain from experimenting with drugs and from having premarital sex. I can only guess that Ron and Nancy somehow thought that sex and drugs were both "bad," but the "Just Say No" campaign had the unintentional effect of making drug use sexy and sexuality druggy.

In keeping with the fine tradition of idiotic fear-mongering bombast, Reagan is quoted as saying: "I now have absolute proof that smoking even one marijuana cigarette is equal in brain damage to being on Bikini Island during an H-bomb blast."

If that's true, he was getting some seriously dank weed.

But it's not just Republicans who stand resolute in the war on drugs. President William Jefferson Clinton, a man famous for not knowing how to smoke a joint, was tough on drugs and in 1998 signed an amendment of the Higher Education Act of 1965—Section 484, subsection R—that took federal financial aid away from college students who'd been arrested for smoking pot. Even President Obama, a man who once said "I inhaled; that was the point," has instructed his Justice Department to get tough on cannabis users, even in states where medical marijuana is legal.

On the walk back to the farmhouse, with the dog leading the way, his farts trumpeting our return like a medieval herald, I chatted with Slim about his upcoming trip to Amsterdam. It was going to be the first time he'd been out of the country and, for Natasha, her first trip on an airplane.

Crockett walked ahead of us, and I could see him running through a mental checklist of all the things he had to take care of. The carefree days of summer were past, the harvest was in, and he had dozens of employees to keep track of. The different strains had to be separated and properly cured, he had to keep track of how much bud they were getting off each plant, he had dispensaries to call and deliveries to make, and he was working on getting his Private Reserve seed business off the ground— trying to figure out how to make his strain a brand name. He was simultaneously production manager, administrator, salesman, marketing executive, and scientist. In other words, being a pot farmer can turn out to be way more stressful than you'd think.

It reminded me of something that happened on the return flight from my first trip to Amsterdam. Crockett and I were standing in the back galley chatting. He'd dropped the pretense that he was "in construction," and we were joined by a young

Asian man who'd also just been to his first Cannabis Cup. He was young and scrawny, looking like a skateboarder with a high SAT score. Crockett and I both raised our eyebrows when we heard that his grandmother had accompanied him to the Cup, but he didn't seem embarrassed or bothered by it at all, treating the whole trip to Amsterdam like a high school student on a college tour. He said he wasn't much of a smoker; he was just checking it out to see if it might be a good career move. But when he told Crockett that growing marijuana looked "easy," I saw a shudder run through his body. To his credit, he could've been sarcastic or dismissive, but instead, a look of sympathy crossed Crockett's face.

"Have you ever grown anything before?"

The young man blinked and shook his head.

"A tomato in your backyard?"

"No."

Crockett nodded. "Do you have any houseplants?"

The young man swallowed. He obviously didn't like where this conversation was heading. "Not really."

Crockett leaned forward, his body dwarfing the young man's, his voice dropping low so no one would overhear. "So you've never had to think about germination, light cycles, watering times, air temperature, CO_2 levels, fertilizer, pests, molds, anything like that?"

Crockett wasn't trying to be mean. He was giving the kid a glimpse into what it takes to grow high-quality cannabis. The young man gulped. "No."

Crockett smiled. "Well . . . good luck with it."

Jerry had returned—his dogs swirling around him in a canine cyclone—and he sauntered over to see what the verdict was on the plants still out in the field. Crockett's cell phone rang and he answered it, drifting away from us to have his conversation in private. He had been waiting for a call from the Humboldt trimmers and was now making arrangements to get them to come down for the weekend. Slim went into the farmhouse to check on his wife.

I can't remember exactly how we got on the topic—but Jerry

started telling a story about being one of the founding members of the Peace and Freedom Party.

"Back in the day you had to smoke dope to be a member." He stroked his beard. "Maybe you still do."

Jerry looked up at me, his eyes magnified by his glasses so that they seemed to have a life of their own. "Of course, I joined for the pussy. The Peace and Freedom Party was all about the pussy."

Crockett hung up his call and lit a cigarette, and Jerry went over to join him. I watched as they walked off toward the trim room, engaged in a private conversation.

Red came skipping out of the farmhouse. He was carrying a pair of knee pads and giggling fiercely, his face flushing in glee, as he entered the trim room to pull some kind of pterodactyl-related prank. I was left standing in front of the farmhouse.

It was suddenly very quiet in the mountains, the chatter of conversation, the inane chuckles of *Beavis and Butt-Head*, the flinty snip of the scissors all faded away. I stood alone in a golden meadow and listened to the plaintive cries of a raven echoing in the tall pines and the gentle sound of the wind whistling out of the dog's ass.

Chapter Seventeen

The Green Rush

"Welcome to Oaksterdam!"

With that, the young woman flashed me a big toothy smile and handed me my Oaksterdam University student ID card. I thanked her and joined a long line of students shuffling up the stairs for the start of the "Basic Seminar"—a two-day intensive that's the prerequisite for all other classes at the university. Maybe it was because it was the first day of school, but there was a hint of apprehension in the air, a kind of free-floating anxiety among the students. No one was talking, really. Perhaps because none of us knew what we were getting into.

Sixteen states and the District of Columbia have effective medical marijuana laws on their books while many more recognize marijuana's medicinal value. This sea change in state laws has created what Oakland's *East Bay Express* calls the new "Green Rush." Just like in the Gold Rush of 1849, Northern California has become the place to go.

At the epicenter of this mad dash for the cannabis dollar is Oaksterdam University, a logical starting point for entrepreneurs and risk takers set on striking it rich in the brave new world of medical marijuana. It's an entire university devoted to giving aspiring growers, dispensary owners, edible bakers, and budtenders a thorough education in all aspects of the medical marijuana industry. Can you imagine a university devoted to, say, tulips?

Oaksterdam University offers "Quality Training for the Cannabis Industry" and is just one of founder Richard Lee's audacious cannabis fantasies turned into reality. In addition to Oaksterdam University, Richard owns a medical marijuana dispensary called Coffeeshop Blue Sky, a cannabis nursery, a hemp products store, and a gift shop. He's single-handedly turned a near derelict downtown into a mash-up of Oakland and Amsterdam that's now known as "Oaksterdam."

Despite being confined to a wheelchair—he was paralyzed in a freak accident while working on the lighting rig for an Aerosmith concert—Richard projects a soft-spoken confidence and vitality. He's in his late forties but looks like he should be in an indie rock band instead of running a cannabis empire. Perhaps it's because he grew up in Texas, but he seems to have an innate ability to surprise the opposition, to choose the bold and audacious move over the safe and politically calibrated positioning preferred by most politicians. It's a kind of stand-and-fight, "Remember the Alamo," instinct. And it's made him a force to be reckoned with in the medical marijuana industry in California.

Richard moved to Oakland in 1997 and began growing medical cannabis in a warehouse. When the state started cracking down on dispensaries and growers, instead of going low-profile, he jumped into local politics and succeeded in getting the Oakland City Council to approve an initiative to make marijuana the police department's lowest enforcement priority. Over time, cannabis has become quasi-legal in downtown Oakland, helping to rejuvenate the city by creating jobs and providing a healthy cash flow from taxes paid by cannabis-based businesses.

Unlike many cannabis activists, Richard eschews the non-profit hippie credo that's been the guiding force behind the movement since the 1960s. He favors a more capitalist approach. He understands that weed is big business and he's not afraid to admit it.

Riding a wave of marijuana-friendly legislation and the success of Oaksterdam, Richard took the next logical step and wrote and financed Proposition 19—the Regulate, Tax, and Control Cannabis Act of 2010—getting it on the ballot in California and thrusting the legalization debate into the center of the political conversation.

According to news reports, he sunk $1.5 million of his own

money into the project. That seems like a big roll of the dice, but compared to the $160 million that failed gubernatorial candidate and eBay bazillionaire Meg Whitman spent, Richard got considerably more bang for his buck.

Not many politicians or pundits gave Prop 19 much thought. It was considered just another crackpot proposal in an already screwy slate of legislative initiatives. But then something changed. The response from the public surprised everyone, including the public themselves. Early polling showed Prop 19 leading, with more than 50 percent of Californians saying that the time had come for marijuana to be legalized. Those numbers were bolstered by a surprising number of endorsements: former police chiefs from Seattle, San Diego, and San Jose came out in favor of ending the war on drugs; Dr. Joycelyn Elders, a former U.S. surgeon general, supported it; as did the ACLU, NAACP, and a mishmash of retired politicians, pop stars, actors, and progressive religious leaders. The *New York Times* even ran a supportive editorial by two-time Pulitzer Prize winner Nicholas D. Kristof.

But despite leading in the polls and having a raft of impressive endorsements—even counting an eleventh hour donation of $1 million from liberal philanthropist George Soros—Prop 19 remained the most underfunded and misunderstood initiative on the 2010 ballot and ultimately lost by just 7 points. That meant that in an election with record low voter turnout, California came within seven hundred thousand votes of ending its war on cannabis.

I was joined on the stairs by Joe, a retiree from Long Island, New York, who looked like an older, rumpled version of George Clooney and spoke with the thick accent of a native New Yorker. In front of Joe was a smartly dressed young woman from Austin who could have been on her way to some sort of corporate presentation. She was friendly, like a lot of Texans, and said she was "stationed" in Texas but was currently "deployed" to Arizona. Joe and I exchanged a look. Her choice of words struck me as a little peculiar. Aren't soldiers and federal agents "stationed" and "deployed"? I can't imagine an assistant regional manager for Pepsi being deployed anywhere. Was I being paranoid or was someone from law enforcement infiltrating the class?

The university is in a converted office building. It's maybe six or seven stories tall, with large and small classrooms, administrative offices, and horticulture labs. The walls are painted white, either to simulate a scientific institution or to save money, and the Oaksterdam logo is splashed around in the garish green and gold of the Oakland A's baseball team. The official "Oaksterdam crest" is a cheeky pun on the famous crest adorning Harvard University, only instead of "veritas" spelled out on top of a background of open books, the word "cannabis" is substituted. One person's truth is another's medicinal herb.

After a second check of our IDs we were herded into a large open room. I found a seat, placed my notebook and number two pencils at the ready, and then checked out my fellow students. Ruth, a thin woman in her fifties with oversized eyeglasses and a stylish beret perched rakishly on her head, turned to me and said, "Look at all the seniors."

She wasn't kidding. There were about 150 people attending the seminar, and more than half of them appeared to be over sixty-five. Does the AARP offer a discount on Oaksterdam classes? Or is this what politicians mean when they talk about privatizing Social Security?

The remaining students represented diverse ages, styles, and interests, ranging from skatepunk stoners looking for grow tips to sport coat–clad businessmen on the hunt for a good investment. Surprisingly, not all of the students were smokers. I spoke to one man, a muscle-bound forty-year-old, who said, "I don't use it. I'm just looking for something to get into." This struck me as a weird thing to admit at a cannabis university. If I was investing in a restaurant I might want to taste the food.

Yet despite these differences, the student body was predominantly male and Caucasian, collegial and attentive.

And there was a lot to pay attention to. The lectures came fast and furious. Day one began with the "Politics and History of Cannabis" taught by Chris Conrad, a "court-qualified expert witness" on cannabis and the publisher of the *West Coast Leaf* newspaper. Chris looked a lot like a classic version of the affable, ponytailed college professor—the one the female grad students always seemed to have an affair with—and somehow managed to cover eight thousand years of cannabis history—including prohibition and the new activism—in ninety minutes.

There was a break and students mingled. I met a couple of guys from Florida, a woman and her husband from San Diego, someone from Rhode Island, and three guys who came together from Colorado. People were flying in from all over the country to take the classes, and there was excitement in the air, like something bigger than ourselves was happening.

I got a cup of coffee and wondered briefly if the invention of powdered dairy-free "creamer" could qualify as a crime against humanity. I surveyed the platters of pastries laid out for the students and ultimately decided that my sedentary lifestyle and buttery croissants were incompatible.

The next lecture, "Legal 101," was taught by James Silva, a criminal defense lawyer who specialized in cannabis and medical marijuana law. Silva could've easily won the Best Dressed Award for his fashionable mocha-colored suit, lavender shirt, and yellow tie. With his suit, swarthy good looks, and Mephistopheles-style goatee he reminded me of Dr. Orpheus from *The Venture Brothers* cartoon. Silva's eyes flashed with intensity as he spoke. He was, not surprising for a trial lawyer, extremely articulate and intelligent. Unfortunately, the information he was so eloquently declaiming was not what any of the students wanted to hear. Whatever hopes and dreams, good vibes, and excitement they might've started the day with were quickly crushed by Silva's reality check.

He addressed all aspects of state, local, and federal law, and the take-away was depressing. Even if you had all the licenses and permits, permissions, and tax ID numbers available through your state, if the feds decided to bust you, well, you were fucked. If you believed what Silva was saying, and I did, it seemed that opening a dispensary was the express lane to incarceration. By the time his talk was over, Silva had pretty much scared the shit out of everyone present, myself included, and I wasn't even planning on growing weed or opening a dispensary.

A woman next to me hung her head and heaved a sigh. She had told me at the break that she was hoping to open a dispensary; now she wasn't so sure. "I don't want to go to jail," she said.

It was time for lunch, and people shuffled out with looks of genuine anguish and despair on their faces. Now that they understood what they were up against, many of them were left wondering why they had ever thought getting in on the "Green Rush" was a good idea.

But the threat of criminal penalties and federal mandatory minimum sentences don't seem to be stopping the proliferation of educational institutions like Oaksterdam University. In just the past few years schools teaching the ins and outs of cannabis cultivation and dispensary practices have sprouted up all over the country—schools such as MedGrow Cannabis College in Detroit and Greenway University in Colorado, which is the first state-approved and -licensed medical marijuana educational facility in the country.

Cannabis activists and cannabis users want the medical establishment to start taking cannabis seriously as a useful treatment. But for that to happen, the federal government would have to allow clinical trials, and for clinical trials to be taken seriously, the people who grow cannabis for medical use need to start developing some quality protocols and controls that keep unscrupulous growers from spraying pesticide or fungicide on plants. It won't help anybody if the meds are tainted. And that's where these universities come in. They're trying to develop safe practices and standards for growers, budtenders, and the medical marijuana industry in general.

During the lunch break I walked over to check out Richard Lee's Coffeeshop Blue Sky, a tiny hole-in-the-wall dispensary just down the street from Oaksterdam. Not surprisingly, a

Grateful Dead song was blaring as I entered. Brightly painted and friendly, Coffeeshop Blue Sky looked like every juice bar you see in a typical college town, only with the added bonus of weed for sale. I showed my doctor's recommendation to the doorman and he pointed me to the back room where the dispensary was.

Imagine opening a marijuana dispensary in your closet and you might get a good idea of what this was like. There were no shelves of glass jars filled with bud, no display cases, no digital menus flashing on the wall. It was just one of those half doors, the kind common in movies about farmers, with a friendly young budtender standing in a space the size of a broom closet. The menu was a three-ring binder that you could flip through. The selection wasn't great—they were mostly the strains that Oaksterdam sold through their nursery—so I purchased a bag of sour cherry White Widow gum drops just in case the afternoon lectures dragged.

After lunch we were treated to "Civics 101," which might as well have been subtitled "Don't Be an Asshole." It was a review of commonsense practices, such as "Be a good neighbor," "Don't smoke in public or around children," and "Be polite and exercise discretion." Oaksterdam, like the Berkeley Patients Group and others, is working hard to shatter the stereotype of the long-haired slacker who sits on the couch taking bong hits and watching cartoons all day. They want to show that cannabis users are vital, active, and intelligent members of society—which, for the most part, they are.

The rest of the afternoon was devoted to "Horticulture 101." Like the other lectures, this covered a lot of ground as fast as possible. It was presented by a local grower named Chris McCatheran, a young black man with a big smile, a shaved head, and all the charm of a marine drill instructor.

The first thing you need to know about setting up your indoor grow room is that it is illegal to booby-trap it. That's right: land mines, tiger traps, punji stakes, bottles of hydrochloric acid balanced above the door, and shotguns with strings tied around the triggers and connected to doorknobs in some kind of Rube

Goldbergian construction are all no-nos. Who knew you could get into so much trouble with a booby trap?

After that introduction, McCatheran blasted his way through the "Oaksterdam method" of setting up an indoor garden and nursing your plants from seedlings to vegetative stage to flowering. The information was technical, and many in the audience were having trouble following. One student raised his hand and asked, "When do you begin the flowering stage?"

McCatheran nodded and replied, "When the plant is half the size you want it to be at the end."

This statement was met with bafflement. The students acted like it was some kind of Zen koan.

Another hand popped up. "How high would you say half the size is?"

"That depends on how tall you want the plant in the flowering stage."

Another hand. "But how do we know when the plant is half as tall?"

McCatheran blinked out at the students and, for a brief second, I thought he was going to start shouting and make us do push-ups.

"That depends on how tall you want the plants to be."

The audience still wasn't getting it. More hands shot up, more questions were asked, until finally McCatheran said, "About four feet tall. Okay?"

There was something definitive in the way he said "Okay?" As in, don't ask any more fucking questions.

As people frantically scribbled the information down, McCatheran looked out at us and shook his head. Were we trying to piss him off?

"Let's hold the questions until we get to the end. We've got a lot of material to cover."

We did have a lot to cover and, to his credit, he covered it. A student sitting behind me turned to his friend and said, "I can't believe growing pot is so complicated."

I now felt that, with my notes, I could build an indoor garden and grow myself some weed. But I was surprised that a lot of the techniques Oaksterdam recommended are different from the ones used by the professional growers I've met. There wasn't any radical difference, but it seemed as if the Oaksterdam formula was created to deliver foolproof medical marijuana for dispensaries, not necessarily to push the boundaries of what the plants could do or to create something dank.

The next day was more of the same, only different. There were classes on making extracts and hash, a lively cooking-with-cannabis lecture, a fascinating discussion of the science and politics behind FDA approval led by Paul Armentano, the deputy director of NORML and an expert in the field of pharmacology, and a rah-rah speech about becoming engaged cannabis activists.

In the two days of instruction I attended at Oaksterdam, the word "dank" wasn't mentioned once. There was no discussion of varietals or the differences between indicas and sativas. And most of the students I spoke with had never even tasted a pure sativa. But maybe that material is covered in the advanced seminar.

In order to graduate from Oaksterdam, students are required to take an SAT-like exam. The instructor said the test would take five hours, but I didn't really believe him. Still, I wanted to do well. I was even hoping I'd snag class valedictorian.

Maybe it was because I hadn't taken a test in a long, long time, but it was much harder than I'd anticipated, the questions were specific and annoyingly technical, and it took me more than three hours to complete. Still, I was confident that I'd done okay and mailed my answer sheet back to the university.

A few weeks later I got the results back and learned that I'd passed. I'd even done pretty well, but I was no valedictorian. Still, I was proud of my freshly printed Oaksterdam diploma. I had learned a lot. But had I learned anything that would help me discern dankness when I went back to Amsterdam and the Cannabis Cup?

For me the highlight of my weekend at Oaksterdam was Richard Lee's closing comments about the current political environment and where he thought the cannabis industry was headed. He's a forward-thinking guy and managed to be both upbeat optimistic and downbeat realistic. The world of cannabis is simultaneously hopeful and fraught with danger. He ended his remarks with some of the best advice I've ever heard: "Keep your head down. Zig zag. Watch your back."

Chapter Eighteen

The Rumble in the Lowlands

It had been one year, almost to the day, since I had taken my first flight to Amsterdam, and here I was again, waiting in the departure terminal at LAX. KLM Royal Dutch Airlines offers nonstop service from Los Angeles to Amsterdam, and the flight the day before the Cannabis Cup was filled with stoners. There were a lot of dudes with beanies pulled down over their heads wearing puffy skateboarder coats, a cute young woman in a pink T-shirt that said "I ♥ Oaksterdam," and a collection of scruffy guys with variations of Bob Marley or Che Guevara T-shirts and hoodies. And then there were the cannabis industry types who stealthily pass through both worlds. Don from DNA was on the flight. So was Doug from the Gourmet Green Room. We were all heading to Amsterdam for the premier marijuana tasting event in the world.

The *High Times* Cannabis Cup consists of two distinct parts: One is the "coffeeshop crawl," which stretches out all across the city, encompassing almost forty-five different coffeeshops; and the other is the expo, similar to the industry trade shows that happen in Toronto and Los Angeles, which is held in a massive nightclub called the Powerzone.

The Powerzone isn't near much of anything interesting. It's located in a somewhat forlorn and desolate neighborhood on the south side of the city, smack in the middle of a marsh. The surrounding wetlands are bisected by highways and dotted with isolated corporate compounds that look a lot like the headquarters of some evil conglomerate you'd see in a James Bond movie. It reminded me of the suburbs outside of Dallas.

To reach the Powerzone, you take a short metro ride from central Amsterdam to the Spaklerweg station. I don't know why, but I love that word. If I had a band, I might call it Spaklerweg. We would probably play something called "ambient heavy metal."

It was a short walk from the station—past a few corporate security kiosks designed to thwart industrial espionage—until I reached a bizarre strip mall. If you can imagine a mall plopped into the middle of the Everglades, then you kind of get the idea. The businesses in the mall apparently don't see much foot traffic and try to catch the eyes of people zooming by on the roads and rails with some of the biggest signs I've ever seen. The sign on the carpet store had lettering large enough to be visible from a low earth orbit.

In direct contrast to the gargantuan signage was a small sloppily handwritten sign that said "Cannabis Cup" tied to a chainlink fence with a series of trash bag twist ties. An arrow pointed in the general direction, while a couple of randomly scrawled hearts promised that warmth, affection, and camaraderie were just around the corner. This was encouraging, because I felt like I was walking toward an industrial loading dock.

I continued through the parking lot around the back of the carpet store, past some trucks and delivery vans, to where another handwritten sign gave further instructions. I turned left and followed a small driveway running in front of an abandoned factory, and found the entrance. A half dozen stoners stood in a circle in front of the Powerzone. They seemed dazed, but I couldn't tell if they'd already been inside or were just arriving.

An old school bus, looking a little bit like the dolphin-and-unicorn-themed RV that Woody Harrelson uses as his dressing room on film sets, was parked in front of the club. A sign stuck in the window indicated that this was the VIP lounge for members of the Temple Dragons. Here was where the secret society gathered, the clandestine lair of arbiters of cannabis excellence.

Before the Cup began I had asked Steven Hager, creative director of *High Times* and the man who started the Cannabis Cup in 1987, if I could observe some of the judging—perhaps be like one of those U.N. election monitors who stand quietly on the side and make sure proper protocols are followed. I also wanted to see the Temple Dragons perform their "420 ritual." Every day of the Cup at precisely 4:20 p.m. the Dragons convene and light seven candles. The candles represent the seven prongs of the cannabis leaf and the seven powers of cannabis: utility, sexuality, medicine, love, poetry, vision, and spirit. According to reports, the ritual is a cross between a Quaker meeting, where everyone is allowed to speak freely, a planning session, and a pagan ritual.

Steve politely rebuffed me. This wasn't a mission for civilians or a place for amateurs. *It was for Temple Dragons.* He made it clear that I wasn't, and probably never would be, a Temple Dragon. Transparency in the judging process is, apparently, not one of their main concerns and is perhaps why accusations of behind-the-scenes wheeling and dealing and outright corruption seem to haunt the competition.

The first thing that hits you as you enter the Powerzone is the skunky fresh scent of high-quality herb being consumed. This is, after all, a venue where competing companies offer samples of their strains, and it's relatively easy to walk from booth to booth enjoying a horizontal tasting—for as long as you can remain physically upright—of some of the best weed in the world.

Just like at the expo in Toronto, Green House had the biggest booth. This time it had a multilevel layout with the upper level dedicated to selling merchandise and a large table below where Davide—Franco's friend and high priest of the chillum—was manning the vaporizer. He was generating massive vapor bags of Super Lemon Haze—the bags were ten or twelve feet long and drew a crowd—and then offered samples to whoever passed by. Arjan and Franco were serious about winning this year. That would make Super Lemon Haze the three-time champ, an unprecedented achievement at the Cup and kind of like winning the Triple Crown in horse racing. It would also guarantee that Green House Seeds would retain its domination of market share.

Their competitors had other ideas.

Barney's Seeds, part of the Barney's coffeeshop/seed empire and Green House's main competition, was gunning for the top spot with a new strain called Tangerine Dream, a combination of G-13 Haze and Neville's A5 Haze. Like Super Lemon Haze, it's a fruit-flavored Haze.

Despite their reluctance to talk about their plans for the Cup, Don and Aaron were stepping strong with six different entries. From DNA Genetics they'd entered two of their best strains, Chocolope and L.A. Confidential, in the sativa and indica cups, and from the newly formed DNA CA—the California branch of the company—they brought different cuts of Chocolope and L.A. Confidential as well as a new strain called Mango OG. Their Reserva Privada line—which promotes boutique West Coast growers and genetics—had entered Kosher Kush and Sour Kush.

And those were just for the seed company cups. There were several coffeeshops entering DNA Genetics strains in the coffeeshop competition as well. A shop called Homegrown Fantasy had entered Chocolope, Green Place had L.A. Cheese, and four other shops had submitted DNA-based entries.

And, as Don said, "There's always the random people who entered it and don't say nothing. Or don't tell us. They grew a cut and just want to put their coffeeshop's name on it."

I think about the time Disney threatened a lawsuit against little Mexican popsicle carts in Los Angeles that had unlicensed paintings of Mickey Mouse on them. "Doesn't that bother you?"

Don didn't look all that bothered. "Whatever, bro. It's all good."

While Aaron organized T-shirts in the back of the booth, Don stood in front, bouncing on the balls of his feet, working the crowd while eating a burrito.

I was amazed that they managed to get so many quality entries together so quickly. For two guys who swore they didn't have a clue what they were going to do a couple of weeks ago, they'd quickly turned things around. Or maybe they weren't telling me the whole story to begin with.

"I thought you guys didn't know what you were entering?"

Don laughed. "We never know until the end, Mark, 'cause we wait until it's the best shit. We put all our shit on the table and we smoke every one. Because you never know, crop to crop."

"Are you going to beat Super Lemon Haze?"

Don shook his head. "They're going to get the coffeeshop cup. And they can have it. We don't care about that one. We don't have a coffeeshop so it's pointless for us. The seed cup is the one."

For a lot of the botanists working in the industry, the Cannabis Cup is more like a popularity contest judged by the general public, where the seed-cup competition—for best indica and best sativa strains—is judged by connoisseurs: the Temple Dragons and celebrity judges.

"I mean, it depends on the celebrities and what they like, but you know, that's what I like about it. You can't buy them. You can't give 'em T-shirts and gift bags."

I wanted to know about the Kosher Kush. It seemed like such a random name.

Don laughed. "Okay, so check it out. There was a strain going around in L.A. that was grown by some Jewish kids. They were calling it the 'Jew Gold.'"

I must've made a face because he held up his hands.

"No lie, bro. And I couldn't call it that. I'm a Catholic kid, my partner's Jewish, and the bottom line is I won't call it 'Jew Gold.' So we called it 'Kosher Kush.'"

"Is it really kosher?"

Don nodded. "A really good friend of ours, this woman, is an ordained rabbi and she blessed it. We're going to stamp it on the packs and everything. It's going to be the first official kosher marijuana strain."

He thought about that for a second, then added, "I don't know if there's some organization or rule book that governs what they allow but, yeah, it's the real thing."

"That would be the Torah."

Don cracked up, a big grin spreading across his face. "Then everything's kosher."

Aaron came over to ask me a question. "Hey, have you seen Frank?"

He meant Franco, from Green House.

"He's at his booth."

Aaron pulled a wad of weed wrapped in plastic out of his backpack. He tenderly unrolled it to reveal a stem studded with seven or eight expertly manicured buds. All the leaves had been taken off but the branch was still intact and the buds were perfect, like an ikebana arrangement in miniature. It was beautiful.

"Cannalope Haze. One of Frank's favorite strains."

This remarkable stalk of Cannalope Haze was a gift from Don and Aaron to celebrate the recent birth of Franco's second son. I followed them over to where Franco was getting the Green House team organized for the day and watched them present the branch to him. In some ways it reminded me of how soccer teams trade their team banners with each other before a match. It's a sign of respect and sportsmanship, and what Don and Aaron were doing was no different.

Franco was obviously moved by their gift. He took a big sniff of the buds and grinned. "Too cool."

Bear hugs were exchanged and then talk turned to the particulars of the Cannalope Haze. Franco could tell, just from smelling it, that it had been grown in organic soil. Aaron thought the buds needed a little more curing before they would be perfect to smoke.

I turned away and surveyed the crowd.

The expo hall was starting to fill up. The big coffeeshops provided free shuttles to and from the Powerzone, and more and more people had arrived.

Franco patted me on the shoulder and beamed.

"Have you seen all the Italians? It's like the Italian Cup, there are so many here."

I could tell that there were more Europeans than at last year's Cup, but in the sea of hoodies they were recognizable only by the languages coming out of their mouths. I heard Spanish, Italian, German, Portuguese, Russian, and some kind of unidentifiable Slavic tongue.

But it wasn't just members of the European Union; there were South Americans, Africans, and hipsters from Asia milling around the expo as well. My favorites were a sharply stylish Japanese couple in chic '50s-style suits who looked like they'd just stepped out of a time machine, and a middle-aged Chinese businessman wearing thick horn-rimmed glasses. The Chinese businessman seemed to be scrutinizing the smallest details of every single booth. Would we see a counterfeit Cannabis Cup in Shanghai the following year?

Naturally, there were hundreds of North Americans. I went to get a beer and ended up next to a guy from Michigan, a beef-fed man who looked more like a high school wrestling coach than someone you'd find at a cannabis convention. He was researching strains that might grow well in a colder, northern climate.

He sipped his beer, a glazed look on his face, and said, "Yeah. We got some good laws in Michigan."

Overall, the crowd seemed younger, more funster fans of the herb than the older, cannabis-industry pros that I had seen the previous year. Maybe that's why the expo felt livelier.

One of the things I find fascinating about cannabis culture at the Cup is the clash between the old-school hippies represented by *High Times* magazine and the younger, hipper smokers who relate more to *Skunk* magazine and websites like Hempista.com, Leafly.com, and BakedLife.com. These are part of a new wave of cannabis-centric sites that mix fashion, design, and pop culture with a refreshingly cosmopolitan point of view.

I have a lot of respect for *High Times* and what they've accomplished. They got the ball rolling and kept it rolling during the dark ages of Reagan's "Just Say No" campaign and the ongoing war on drugs. They have promoted connoisseur-quality marijuana and the progress made by growers and strain developers over the years. I'm not sure that there would be a booming international seed industry without *High Times*. And, let's give them their props: They started the Cannabis Cup. And yet, you can feel that change is in the air. *High Times* takes a proprietary view of the culture that seems a little narrow, a little stale, in an Internet-savvy world. Women, from riot grrls to urban professionals, are becoming growers and activists and are starting to make their voices heard in what has traditionally been a male-dominated industry. Today *High Times,* where too many of the ads rely on scantily clad women frolicking among the buds, seems decidedly old-fashioned, the design and content locked in the zeitgeist of the 1970s. I think the counterculture is more diverse and sophisticated than *High Times* gives it credit for.

This is not to say that *High Times* doesn't have some excellent writers, important horticulture information, and thought-provoking political content. David Bienenstock, Nico Escondido, Bobby Black, and Danny Danko are literate and articulate observers of the cannabis scene, and Jorge Cervantes has positioned himself as one of the leading experts on marijuana growing. They just need to ease up on the tie-dye.

A good example of this hippie versus hipster schism was illustrated by the grumbling I heard from the older generation about the music programming at the nightly concerts and parties. These are a large part of the Cup's appeal, and this year the focus was "Old School Hip-Hop." The gray-haired, progressive-lens contingent wasn't happy about listening to rap music. They wondered why bands such as Quicksilver Messenger Service, Moby Grape, and Jefferson Starship weren't playing. I actually had someone say to me, "I wish Quicksilver were here."

In the Cannabis Cup program, cup director Steven Hager explained the choice of hip-hop by linking the music to blues, jazz, and punk rock, correctly placing it in the continuum of the counterculture movement. But while the hipsters, college kids, and Europeans were digging the beats and rhymes of Dilated Peoples and Del the Funky Homosapien, the hippies I talked to weren't buying it. As one baffled old-timer said to me, "How can you have a Cannabis Cup without reggae?"

It didn't take long before I found myself at the booth of a Los Angeles seed company called the Cali Connection. The previous year they had had a small table and were a relatively unknown upstart, but their business was starting to take off thanks to distribution through DNA's Reserva Privada line and the quality of their strains. In fact, Danny Danko of *High Times* named two of Cali Connection's strains in his roundup of the "Top Ten Kush of 2010."

The Cali Connection is the brainchild of an underground botanist named Swerve, a wiry young dude from the San Fernando Valley whose face seems swallowed by his scraggly beard and the baseball cap that he wears at all times. He's simultaneously scruffy and nerdy and cool—the kind of guy you might see playing analog synthesizers in an alt rock band. Swerve's rise in the industry had been steady, but he hadn't yet broken in to the pot-smoking public's consciousness. A Cannabis Cup win would change all that and that's why he was here.

The Cali Connection is known for extremely potent indica-dominant strains like Tahoe OG, Larry OG, and SFV OG. I asked Swerve what he was entering in the competition.

"We entered Blackwater for indica and Jamaican-Me-Crazy for sativa."

I laughed. "Jamaican-Me-Crazy?"

Swerve scratched his beard and nodded.

"Yeah, it originated in Kingston, Jamaica. We got it to Cali and it's a really fast-flowering sativa. It finishes in nine weeks solid and it's as sativa as it gets. It's the weirdest thing. I've never seen anything like it."

"Anything in the coffeeshop-sponsored competition?"

Swerve shook his head. "We didn't do coffeeshops this year because we were very ill prepared for it. We had a couple of bad runs, nothing really panned out."

Swerve shrugged apologetically. "I have too high a standard."

One of the things I like about the Cali Connection is their logo. I know it sounds weird, but they use a Mario Puzo–inspired image and when you click on their website, you get a needle drop of the theme from *The Godfather*. Compared to the slick corporate style of Green House and Barney's, or the street art cool of DNA Genetics, Swerve has concocted a look that is distinct, iconoclastic, and appealingly goofy.

"What's up with the Mario Puzo–looking logo?" I asked.

Swerve looked at me like I was an ignoramus. "I'm Italian."

. . .

I knew that Doug, the tub-thumping vapemaster of the Gourmet Green Room dispensary in West Los Angeles, was coming to the Cannabis Cup—we'd been on the same flight—so I wasn't surprised to see him stroll into the expo. I could tell he was excited to be there. He was practically skipping around the booths with a huge grin on his face—the proverbial kid in a candy store. I wouldn't have been surprised if he'd suddenly sprung into a cartwheel.

I wanted to sample some of the contenders and Doug was happy to join me.

First up was the reigning champion, Green House Seed's Super Lemon Haze. I took a couple of long hits off the massive vapor bag that one of the Green House employees was wielding. It was the same fantastic citrus flavor that I'd tasted the previous year; the only difference was that this year's version seemed even smoother.

One of the positives about sativa is that it's energizing. You get a kind of chatty rush without the teeth-grinding penchant for ultraviolent mayhem you get with, say, crystal meth. A real sativa delivers a boost that's more like a good shot of espresso.

A few vape hits of Super Lemon Haze and Doug, who's excitable and loquacious by nature, was suddenly orating expansively on a number of subjects including, but not limited to, the differences between Plato and Aristotle and current theories circulating in the psychiatric world about pain and pleasure. The gist of the former was over my head, but the psychological hypothesis suggested that people exposed to a lot of deep psychological and physical pain early in life have a greater capacity for a variety of pleasure as they get older.

I considered my own pain versus pleasure threshold as I went over to the Barney's booth and let an attractive young woman in a red dress shoot freshly vaporized Tangerine Dream down my throat. This strain has an unbelievable flavor, like a big hit of tangerine-flavored Jolly Rancher. It's similar to the orange notes in Jillybean, but where the Jillybean has a delicate flavor, this hits you over the head. Tangerine Dream delivers an upfront sativa wallop with a classic Haze chaser. It was very potent, no doubt about it, and for the first time in a couple years, it looked like there was a strain capable of capturing the public's imagination and giving Super Lemon Haze some competition. Tangerine Dream had the added advantage of sharing a name with a quintessential electronic krautrock band. But I couldn't figure out how it got the tangerine flavor. Did some rare phenotype of G-13 Haze pop up tasting this way? Or was it sprayed with some kind of flavor additive?

I asked one of the Barney's Farm botanists about it. He shrugged and said, disingenuously, "Darwin?" As if natural selection could make a plant taste like a Jolly Rancher.

The Tangerine Dream sent Doug off on a sprawling, multi-tentacled conversation that reminded me of an M. C. Escher drawing. Original, fascinating, with stairs and doorways, corridors and secret passages, spinning off in all directions and dimensions, yet ultimately leading back to the same place. In this case it was something about Aristotle.

Just like when tasting a flight of wines, where you work your way up from a crisp sauvignon blanc to a rich chardonnay to a ruby pinot noir, we finished our tasting flight with something a little more substantial. The DNA Genetics booth was cooking up a giant vape bag of Sleestak. I was familiar with hash made from Sleestak resin, but I'd never smoked it in its original form. DNA hadn't entered the strain in the competition; they were just sampling it because they finally had some Sleestak seeds for sale. Or maybe they were just showing off.

Sleestak is potent, tasty, and pushed me over the edge.

The Sleestak must've jolted Doug's brain, too, because he added a level of athleticism to his conversation, punctuating his points by throwing his body backward and hopping around like a kangaroo-powered philosophy professor in some kind of combination Muay Thai boxing match and grad school lecture. I watched as he hopped off into the crowd at the Powerzone, arms flailing, extolling the virtues of Aristotle.

Crockett, the farmer from the Sierras, had flown to Holland to attend the Cup. He was hunting for seeds, looking for new strains to test out in the spring. I asked him if he was looking for anything specific.

"I like to try a lot of different things. See what works. The great thing about coming here is I actually get to try them before I buy them."

He stood at the bar in the middle of the expo and rolled up a fat joint of Chocolope. He smiled and said, "I haven't smoked anything yet today."

One of the ways Don and Aaron have expanded DNA Genetics is by starting what they call their Reserva Privada line. These are heirloom strains from small producers, boutique growers, and underground botanists that don't have the marketing clout or distribution infrastructure to get their strains out to a wider audience. Crockett has always dreamed of making his Private Reserve available to a wider audience, and I had told Don and Aaron about the unique qualities and historical provenance of his varietal. They were intrigued, so introducing them seemed like an obvious thing to do.

Crockett was a little nervous about meeting Don and Aaron; I think that's why he sucked down the joint faster than he normally would have and it probably explains why he immediately rolled another one and smoked that. But he didn't have to worry. There are a lot of people in the weed world who talk a good line of bullshit—wait, let me rephrase: There are a lot of people in the world who talk a good line of bullshit—but after answering a few questions to establish his bona fides, it quickly became apparent that here was a meeting of like-minded cannabis fanatics who actually knew what they were talking about. It was like watching veterans of the French Foreign Legion share their wartime experiences. They swapped grow room accident horror stories: Aaron once had a pressure sprayer explode in his face, sending him to the hospital; Don has scars from catching hot grow lights before they burned his plants; and Crockett routinely battles poisonous snakes, carnivorous predators, the Mexican cartel, and the DEA. They compared notes on soil and fertilizers and temperature and grow times. They discussed chemical formulations used to make plants produce feminized seeds. In other words, they talked shop.

They must've chatted for an hour, making plans to get

together and work on getting Crockett's Private Reserve seed-making operation up and running and into the Reserva Privada line. This was, to borrow a sports analogy, like having the Dodgers call you into spring training. Don and Aaron shook hands with Crockett and said they'd be in touch. They headed back to their booth. It was the expo, and they had business to conduct. I could tell Crockett was excited, but then he hesitated, as if he wasn't sure if he should celebrate or not. He turned to me.

"You haven't smoked too much. Did what I think happened really happen?"

I nodded. "I think that's what happened."

And then he allowed himself a big Chocolope smile.

Despite the accolades Swerve had earned, the Cali Connection was still an underdog in the Cannabis Cup. But, like Shakespeare once said about blind weed tastings: Every underdog has his day. Without the influence of brand identities and free T-shirts and grinders, it's anybody's game. I'd tasted most of the other big sativa entries. I had my idea of how the awards would shake out, but Swerve's entry was the wild card. The only problem was it wasn't available at any of the coffeeshops in town. If you wanted to taste Jamaican-Me-Crazy you had to be a Temple Dragon.

I decided to take a direct approach. I went up to the Cali Connection booth and asked Swerve if he had any samples of his fast-flowering Jamaican strain. He thought about it for a minute. Stroked his beard. Looked off into space. And then he came to a decision.

"Yeah. I think I've got some."

Swerve knelt down under the booth's table and rummaged around in his backpack. He pulled out a couple of bags of weed, smelled one of them, put that away, and extracted a rumpled plastic baggie with a small, muddy-looking clump of weed in it. He looked up, apologetic.

"It's not what I'd call the best example of the strain. But it'll give you an idea of it."

He sorted through the small nuggets and stuffed a couple of grams worth into a small plastic bag. He stood up and handed it to me.

"Watch out for seeds."

I went in search of someone to roll it up for me.

Smoking Jamaican-Me-Crazy was a strange experience. It had a decent enough flavor—it wasn't fruity or sweet or citrusy. It had an earthy, dirty charm, like chickens ran around in the shade of it, like it was raised in the middle of a tenement yard in Trench Town.

Swerve summed it up like this: "It goes right to your head and stays there. It's too sativa for me. It makes me crazy."

I stood at the bar with trimmer extraordinaire Cletus "The Dingo" McClusky and watched as he rolled some kind of super joint with Cheese—a strong indica—and some Moroccan hash crumbled up in it. I sipped a cold Heineken and smoked about half a joint of Swerve's sativa and waited for the pot to go right to my head, to Jamaican-Me-Crazy, but nothing was happening. This is, of course, a complaint that Californian pot smokers have with pure sativas. They're used to the poleaxing charm of OG Kush, or the cannabis catatonia of a Cheese and hashish spliff, so more often than not, they find the subtle charm of a pure sativa annoying.

I finished the joint and decided it was time for me to get out of the expo and into the streets of Amsterdam.

Of course I started to feel the effects the moment I mounted the steps to catch the elevated train. Jamaican-Me-Crazy builds slowly, over time. It's a real creeper. But when it finally hits, it packs a punch. I was much higher thirty minutes after I smoked it and even higher than that after an hour. This time-release effect probably wouldn't help the strain win in a blind tasting, where its charm would be obliterated by the faster-acting varietals that followed or preceded it.

The sky was filled with dark clouds, and a cold wind whipped through the outdoor station. I stood and waited for the train with a clutch of commuters, enjoying the ominous weather. I heard a low rumble of thunder and then the clouds broke open and poured caviar-sized hail everywhere. Tiny perfect balls of ice pelted down, bouncing off the sidewalk and careening off umbrellas and people's heads. While everyone else ran for the glass shelters, I stood in the hail. It was beautiful, like taking a shower in diamonds.

Chapter Nineteen

A Grateful Dead Reference Emerges in the Narrative

The Grey Area was slammed, packed with people six deep trying to get to the counter. It was always crowded in the tiny shop but this crush was insane. Everyone was there because the coffeeshop has entered one of the most talked about strains in that year's Cup: Casey Jones, a cross of Trainwreck, Thai, and Lemon Haze.

Trainwreck. Casey Jones. Get it? As in the Grateful Dead song with the refrain "Drivin' that train, high on cocaine."

Jon Foster was working the train angle. He had stickers on the bags of bud with a cartoon of a buzzed and befuddled-looking engineer driving an old steam train. He was passing out train whistles with the Grey Area logo; he had little alt-rock pins for your jacket, and he was playing railroad-themed music punctuated by a special-effects train whistle that he blasted every now and then. He was obviously enjoying himself, grinning like a maniacal five-year-old whenever the "train ran through the station" in his tiny shop.

The crush was intimidating, like some kind of fraternity prank to see how many people they could smush into a dorm room. I decided to wait outside. On the street, where a light rain had begun to fall, people stopped, looked in, and then decided to move on to another coffeeshop. Still, the vibe was friendly. There was actually a lively scene taking place out front. I chatted with a German skatepunk and a well-dressed exchange student from Mozambique. We watched as a group of festive Italians took turns posing in front of the window. A couple of very stoned Serbs stumbled out and consulted their map, trying to decide which coffeeshop to crawl to next. I talked to three guys from Florida, young Americans on their first trip to Amsterdam, as they rolled up to the front door and hesitated.

"Looks kinda crowded."

They were, heroically, trying to visit every coffeeshop on the coffeeshop crawl. This year there were thirty-four shops participating and these guys had already hit seventeen of them.

I asked if anything had blown them away.

One of the Americans, a clean-cut young dude wearing a baseball hat with an image of a straight razor on it, stroked his chin thoughtfully. "I liked the Tangerine Dream at Barney's." His buddy, a rotund and friendly-looking fellow with red-rimmed eyes and a Peruvian knit cap scrunched down tight over his head, concurred. "And the triple Lemon Haze from Green House. That one's great."

The Grey Area crowd gave them pause, but they would not be denied. They were going in, one way or another. They pushed through the door, fought their way to the counter, bought a gram of Casey Jones, and popped back outside. There was nowhere for them to sit so they headed off to Amnesia where they hoped to find a quiet spot for a smoke.

A lot of Jon's business at the Cup seemed to be of the "I'll have that to go" variety.

Inside people were standing shoulder to shoulder and hip to hip while they took their bong hits. But nobody was complaining. Jon had his friendly living room vibe going—there were just a whole lot of people in the living room today.

Standing in the middle of the crush was a pudgy American dude with blond dreadlocks that dropped down to his waist like fancy tassels. He was trying to take a rip off a tall glass bong. It reminded me of rush hour on the New York subway—maybe the number 1 uptown local. The doors opened and people squeezed in and out of the shop. This tightly congested influx and outflow buffeted the dreadlocked dude—bumping him one way, pushing him another—but he'd attained some kind of equilibrium. He didn't spill the bong water and actually seemed to be enjoying the ride. Every time Jon hit the train whistle, the dreadlocked dude's face contorted in a classic air guitar grimace and he would raise his hand and pull on an imaginary cord.

Casey Jones, you better watch your speed.

A young black woman with hipster glasses and a perfect afro squeezed past, a burning spliff in her hand, while I shared a table with an attractive couple from Tampa Bay. They were smoking as much as possible before they caught a train to Paris.

Since I'm incapable of rolling a joint, Crockett joined me to check out the strain. I opened the baggie and pulled out the Casey Jones. The bud was gorgeous, which I would expect from one of Jon's selections, and had a pungent pine/citrus scent. It smelled clean. Or maybe it smelled like cleaning fluid. Crockett rolled two perfect joints—I mean, seriously, the dude could work for Marlboro—and we lit up.

Jon had told me that the strain "gets you up and then lays you out pretty good."

I have to admit I had some apprehension. I am not a fan of Trainwreck so I smoked Casey Jones with trepidation. I was hoping that the Thai and Lemon Haze components would somehow counter the Trainwreck.

While Casey Jones didn't give me the soaring uplift or visual thrills of an equatorial sativa, it did have a very nice high. I was buzzed but social, relaxed but not comatose. Even though it had started to rain hard, Crockett and I decided to escape the crush of humanity overfilling Grey Area and find something to drink. Once I stepped outside, the Thai influence—the sativa clarity—was immediately apparent.

We strolled a few blocks and eventually found a bright and modern pizzeria called Mazzo. We sat down and I relaxed, letting the lingering buzz from the pot expand in my head as Crockett told me how he'd been kept awake most of the night by uncontrollable giggling from the next room. He shook his head and said, "Apparently the Guru and Cletus found a smart shop."

Smart shops sell psychedelic mushrooms.

At one point, just so he could sleep, Crockett had wedged a

chair under the doorknob to his bedroom to prevent any mischievous mushroomers from invading his room.

We ordered a couple of cracker-thin pizzas—the sauce seemed to have been painted on with a brush—and shared a bottle of tart Sicilian wine. It was, just like the meal at Lee in Toronto, one of those situational moments that make smoking cannabis so pleasant. The weed was very good, but not overpowering; the restaurant was friendly and comfortable; I was in good company; the pizza was tasty and the white wine delicious. All these factors combined to make the experience of smoking Casey Jones dank in a way that I think Jon Foster would've approved of. I suddenly found myself rooting for Casey Jones, hoping that the strain could break the lock of the corporate sponsors and give Grey Area a Cup win.

Chapter Twenty

The Tribe
Has Spoken

As the big night approached, there were signs that an upset might be in the making. There were ominous rumblings emanating from the "at large" judges I'd talked to on the street. Buzz was building for Tangerine Dream. Of course, people still liked Super Lemon Haze, and Green House had a loyal following, but there was a general feeling that enthusiasm for the two-time champion was waning.

One of the great things about the Cannabis Cup is that anyone who buys a judge's pass can be a judge in the main, coffeeshop-sponsored competition. That means they can vote for their favorite strain, best import hash, and best Dutch hash entered by the various coffeeshops around town. This can be seen as either a true test of a strain's dankness or as a corrupt popularity contest between corporate sponsors. These awards are different from the seed company cups, which are judged by a panel of experts. Interestingly, the 2010 Cannabis Cup was turning out to be different. It wasn't the usual clash between two titans. There were other strains in the mix, strains from coffeeshops that were usually relegated to the sidelines. It seemed like everybody had a favorite, and none of them was from Green House or Barney's.

Traditionally *High Times* releases a "short list" of the best coffeeshop entries chosen by the panel of celebrity judges at the halfway point in the competition. The short list doesn't decide the winner but is supposed to be a guide for the at-large judges, a recommendation pointing them in the direction of what the experts feel are the best entries. The short list is also an early indicator of which strain might ultimately be crowned champion.

This year the celebrity panel included Addison DeNour and Russell Wiggins from Steep Hill Labs in Oakland; Dale Gieringer, the director of California NORML; a few "hip hop legends," including Coke La Rock; and the reigning Miss *High Times,* among others.

The celebrity panel had sampled all the coffeeshop entries and, shockingly, Tangerine Dream, Super Lemon Haze, and Casey Jones hadn't made the list.

Instead, the short list was composed of mostly unheralded strains from small coffeeshops. The panel chose Sleaze from the Noon, Amnesia Haze from Bluebird, Chocolope from Arabica Lounge, Strawberry Kush from the Bush Doctor, Chemdawg from De Kroon, S-5 Haze from Prix D'Ami, and something called Kass Kush from Betty Too as their top picks.

I asked Jon Foster at Grey Area what he thought, and he just smiled and said, "It's the Cup. There's no telling what'll happen."

The awards ceremony was held in a sprawling arts complex and cultural center called the Melkweg. Melkweg translates to "milky way" in English, and it's unclear if the name comes from the fact that the building is a refurbished dairy factory or if it's because of the galaxy of events held there. The nonprofit center houses concert and theater stages, screening rooms, art galleries, a café, and a couple of bars.

The official Cannabis Cup parties and concerts were held at the Melkweg, and every night there was something different going on. On separate evenings, Green House sponsored a party with DJ Muggs from Cypress Hill and Kid Cudi, *High Times* hosted Del the Funky Homosapien, and DNA Genetics threw their annual "Hot Boxxx" party with Dilated Peoples head-lining a wrecking crew of DJs. It's a testament to the fortitude of the Cup-goers that they could crawl through coffeeshops all day, smoke joint after joint of great weed, take bong hits of potent hash, gobble a space cake, and still show up ready to drink and dance until midnight. Perhaps cannabis is some kind of secret performance-enhancing drug.

. . .

The 2010 *High Times* Cannabis Cup awards happened to fall on Thanksgiving, so I began my evening by having dinner at an *eetwinkel* ("food shop") called Het Magazijn, which translates to "The Warehouse" in English. The name is a joke; the restaurant is tiny, smaller than the Grey Area coffeeshop, and has only one table. Eight, maybe ten, people sit together and share a meal prepared by a sardonic chef who seems to specialize in ironic asides as he works in a kitchen a few feet away. The food is very good, kind of Dutch-influenced Italian cuisine. The wine list is a choice between red or white.

Het Magazijn is not pretentious and it's not expensive and they don't make grilled cheese sandwiches, which is why— along with De Knijp ("The Squeeze") and De Waaghals ("The Daredevil")—it's one of my favorite restaurants in Amsterdam.

Outside, it was freezing. By that I mean it was zero degrees centigrade, the temperature at which water solidifies. I reluctantly bundled up and left the cozy warmth of Het Magazijn and strolled along the Stadhouderskade, past the Rijksmuseum, shivering my way toward the Leidseplein and the Melkweg.

The Leidseplein, normally a small square surrounded by bars and restaurants, had been transformed into a winter wonderland. A small ice-skating rink stood where cars normally parked, the trees were festooned with Christmas lights, and a variety of rustic huts had sprung up selling holiday favorites such as piping hot spiced wine, gigantic sausages grilled over an open flame, and, of course, beer. You've got to give Amsterdammers credit because, despite the arctic temperature, the square was crowded with people enjoying the festivities. For me, someone from Los Angeles, there is nothing stranger than seeing a grown man standing in the sleet, his breath forming an icy plume in the frigid night air, hoisting a frosty mug of beer.

I sauntered into the welcoming warmth of the Melkweg, bought a Heineken, planted myself right behind the sound mixer, and waited for the awards ceremony to begin.

A couple of Italian hipsters stood next to me, drinking beer and smoking a spliff. Above their heads was a sign in English that said "Smoking tobacco products is prohibited by law." Smoking cannabis, on the other hand, was not, and almost everyone in the crowd had a joint or a pipe or a pocket vaporizer held up to their lips.

Gigantic close-ups of the buds entered by the various coffeeshops—the samples beautifully lit and rotating like some kind of agri-porn—were projected on a big screen in front of the stage. Partially hidden behind the screen, a chaotic comedy unfolded. The *High Times* staff scurried around, arguing, laughing, looking for things they may or may not have misplaced, and basically acting like no one was really sure who was in charge or what was going on.

As the crowd grew in size, swelling to capacity, the concert hall became seriously smoky. A couple of guys from Boston wedged between me and the Italian hipsters and took out a small pipe and a wad of hash. One held their beers while the other lit up, then the pipe and lighter were exchanged for the beers and the process repeated.

A centerfold-worthy photo of a Tangerine Dream bud appeared on the giant projection screen and a large contingent of Barney's fans began to hoot and cheer with real enthusiasm. When close-ups of Super Lemon Haze appeared, the silence was jarring.

Forty minutes after the awards were scheduled to start, the screen lifted and the emcee began calling for the vendors who'd had booths at the expo and the coffeeshops that had participated in the Cup to "send your representation to the stage" and receive a "participation trophy." It was the "everyone's a winner" moment that you see at youth soccer games all across the United States. *Everybody gets a trophy!*

"Send your representations to the stage!"

Instead of seven-year-olds clutching juice boxes and wedges of freshly sliced oranges, this was a shuffling herd of stoned and exhausted grown-ups moving like a proverbial herd of cats, wandering across the stage to picked up their trophies.

"Send your representations to the stage!"

Next there was the induction of a new Temple Dragon, a young volunteer who'd been working the door of the expo and generally acting as an all-around gopher for the *High Times* staff. After some hugs and the presentation of a spiffy jacket, she lit the seven sacred candles—one for each part of the leaf—and the awards show began to pick up momentum.

The first award presented was for Best Booth at the expo. Considering that most of the booths were just a couple of tables shoved together with some display cases on top, it seemed like a ludicrous award.

Green House, who had shown off their large and beautifully designed display at THC Expose in Los Angeles and at the *Treating Yourself* conference in Toronto, had to deal with the space limitations of the Powerzone, and their booth looked more like a pop-up store at a mall than their normal presentation.

The awards were read in reverse order with Attitude Seed Bank taking third place, Green House Seeds coming in second, and Barney's Farm taking first place.

Franco grinned and waved to the crowd. "Actually, our booth was not so special, so this was a tribute to my crew."

Franco is nothing if not honest.

For connoisseurs of cannabis, the awards that carry the most weight are the seed company categories, especially the Sativa Cup. The Temple Dragons and celebrity judges had sampled all the sativas and, with surprisingly little fanfare, announced the winners. Third place went to Sour Power from a small Dutch seed company called HortiLab, second place went to DNA Genetics for Chocolope, and first place went to an old-school strain, Acapulco Gold, from another small seed company, Amnesia Seeds.

There was a sense of shock in the audience. While it's arguable that most connoisseurs wouldn't award first place to Super Lemon Haze or Tangerine Dream, the fact that neither one of them even managed to place or show was a big surprise.

The guy from Boston nudged me with his elbow.

"Did you try Acapulco Gold?"

I shook my head. I hadn't heard a thing about it.

He grinned. "It was awesome."

From the expression on his face, I believed him.

Don and Aaron came rolling onto the stage with a relaxed, breezy charm, as if accepting awards was just a cool thing they did a couple times a year. This is not to say they were too cool for school. They were sincerely happy, sporting huge grins, laughing with each other as they waved to the crowd.

Aaron walked to the front of the stage and reached into a small box he was carrying. He began tossing handfuls of joints into the crowd as Don picked up the microphone.

"Second place again."

Don smiled at the crowd.

"It's still good, guys. Glad you guys like it."

And with that, he handed the microphone back to the emcee and walked off the stage with Aaron.

Sissi, a zaftig woman with an explosion of black curls bursting from her head, accepted the award for Acapulco Gold. She was excited, bouncing on her feet and hyperventilating into the microphone.

"This means a lot. A Sativa Cup is, you know, the cream of the cream."

She jigged around the stage, holding the trophy as her speech began to ramble, thanking various people and telling a story about an older pothead from the United States who'd tasted the Acapulco Gold and told her, "Hey man, this is really the old-school shit."

That, she said, "made it all worthwhile."

This year the competition added a new seed company category: a Hash Cup. I couldn't quite figure out the logic. What did strain developing and botany have to do with hashish making?

Mila Jansen, the "hash queen of Amsterdam" and inventor of the Pollinator and Ice-O-Lator hashish-producing machines, walked out to present the award. Unlike the majority of the presenters, who wore jeans matched with various logo T-shirts, Mila brought some European style and sophistication to the show by wearing a jacket that was somehow both elegant and psychedelic.

Unfortunately she had lost her voice and was able to make only some indecipherable croaking sounds, as if a frog had suddenly become an expert on the production of Nederhash and hopped on stage. Starbud Melt from HortiLab took first place while Don and Aaron took second and third with Sour Diesel from their Reserva Privada line and Blackberry Kush from DNA Genetics.

Again they sauntered out and tossed joints into the crowd.

A rapper by the name of Curren$y came out to announce the winner of the Indica Cup. He bumped and bumbled his way around the stage like he'd never been on one before, and it soon became apparent that Curren$y was completely baked. At one point he turned to the audience and said, "Yo, am I fuckin' up? Haaaaaaa."

He did, in fact, seem to be fucking up, and *High Times* staff writer Danny Danko came out on stage to try and move things along. The expedited awards were quickly announced. Third place went to White OG from Karma Genetics, second to Cold Creek Kush from T.H.Seeds, and the Indica Cup went to Reserva Privada's Kosher Kush.

Don and Aaron came out and this time their joy was undisguised. They raised their hands in the air like Rocky after he knocked Apollo Creed to the mat. Curren$y looked at them, blinking, as if he'd just remembered he'd seen them somewhere before and said, "You guys are winning everything."

Don grabbed the microphone and exulted. "What a night!"

Aaron stood next to him, holding the trophy, as Don continued his speech.

"This is really special to us. We last won a cup five years ago on this stage. We got our Indica Cup, so now we have one of each."

He was referring to the Sativa Cup won by Martian Mean Green in 2005.

Don handed the microphone to Aaron, who appeared to be genuinely moved by the win. His eyes welled up as he thanked the judges and the fans. "We're West Coast true to the heart and we bring the herb from the West Coast."

Don and Aaron didn't know it at the time, but they were on their way to a big year of kicking ass at cannabis competitions. A few months after their dominant performance in Amsterdam, they were once again walking offstage with more awards than any other seed company at Spannabis with wins for OG #18, Chocolope, and a Sour Diesel hash.

· · ·

The coffeeshop awards followed and seemed somehow anti-climactic. The Nederhash Cup, which goes to coffeeshops selling hashish made in Holland, went one, two, three, to Green House, Barney's, and Grey Area.

Jon Foster thanked the fans for his third-place finish and graciously encouraged them to "enjoy the Cup!"

Arjan and Franco took the stage and Arjan—I think sensing that it wasn't going to be his night—made a speech that was heartfelt and generous and, at the same time, a bit of a fuck-you to the judges who'd snubbed Super Lemon Haze in the Sativa Cup category.

"After winning so many cups, I think this is thirty-six, we want to honor one person with this cup. She's never won one and she's the mother and the grandmother of all the ice and hash here in Holland. So we're gonna donate this cup to Mila. She has been so important to our industry I think we should recognize her."

Mila came out and hugged Franco and Arjan. She was obviously touched by the gesture.

Franco waved to the crowd. "This is special."

The Import Hash Cup was the reverse of the Nederhash Cup with Green House second and Barney's first. A young man with long straight hair accepted the cup for Barney's. He looked strikingly like the hapless lentil-cooking hippie Neil on the old BBC sitcom *The Young Ones* as he meekly mumbled his thanks and wandered off.

That left the top prize to be awarded, the Cannabis Cup, the best overall strain as voted on by the fans/judges who managed to crawl from coffeeshop to coffeeshop and sample as many as they could. As the tension in the room increased, measured only by the thickness of ganja smoke, I found myself torn between the desperate need for another beer and the fact that if I went to the bar, I would lose my prime viewing position. A refreshing Heineken, it seemed, would just have to wait.

Amazingly, the crowd grew quiet, turning their attention away from whatever they were smoking or drinking and toward the stage.

Third place went to the Green Place coffeeshop for L.A. Cheese, a strain collaboration between DNA Genetics and Big Buddha Seeds. Once again, Don and Aaron walked out on the stage to claim a prize. This time they were joined by Big Buddha himself—not *the* Buddha, but a strain developer who goes by the name Big Buddha—and the owner of the Green Place coffeeshop. A quick tally gave Don and Aaron a first-place Indica Cup, second-place Sativa Cup, second and third in the Seed Company Hash category, and—although technically this award went to the coffeeshop—a third place in the big prize.

Second place went to Super Lemon Haze and Green House United coffeeshop. And the grand prize, the Cannabis Cup, was taken by Barney's for Tangerine Dream.

If Arjan was disappointed that Super Lemon Haze didn't make it three in a row, you couldn't tell. He smiled and thanked the judges, before offering his heartfelt congratulations to Barney's. Franco was equally gracious in defeat.

Derry, the owner of Barney's and a man with a passable resemblance to Robert Plant, came to the stage to accept his award. He projected a measured, almost magisterial, air as if the cup was his birthright or, perhaps more accurately, he was the white knight who had crushed the dominance of Green House and wrested the grail from Arjan's grubby hands. Derry held the cup up in the air and accepted the sustained applause of the audience. He bowed a few times and then took the microphone.

"This is an honor like nothing in my life, and I am honored and humbled before you."

It only took a few hours for the haters to begin writing scathing comments about the results on the *High Times* website. Interestingly, the comments had a couple of recurring themes. Most notable was the belief that the Cup was rigged. One anonymous poster said, "Everyone knows the cup is a joke. HT, you'd be better off revamping the entire thing because otherwise you'll soon be flogging a dead horse. Nobody believes anymore that Barneys and Greenhouse consistently do the best weed in Amsterdam. NOBODY!"

Another anonymous commenter agreed: "High Times

Cannabis Cup is more like the Coin Toss Cup Greenhouse V. Barneys"

A lot of commenters thought that Tangerine Dream had been sprayed with some kind of flavor additive. A typical comment was from a writer named Tangerine Spray Dream who said, "spraying the cannabis (amnesia) white orange jus = Tangerine Dream." I take this to mean he or she thought that the winning strain was just orange-flavored Amnesia Haze. Grower 420 agreed, stating, "The tangerine dream and super lemon haze have too mutch fruit juice in the buds to smell and taste like that!"

I've tasted California-grown Super Lemon Haze, and I can say, with certainty, it does taste like that.

A number of people thought that Casey Jones from Grey Area was the best cannabis in the competition, as a commenter called Honest judge made clear. "I guess first and second places are bought! Barney's and the Greenhouse's product this year were not the best weed in Amsterdam just the most freely distributed!! Well done Green Place and Grey Area—Nice weed ;-)."

The more direct response was reflected by this comment from HEKTIC 718: "CASEY JONES@greyAREA SHULDVA WON." This, by the way, I don't disagree with.

Some visitors to the Cup were disappointed by the vibe of the event, as another anonymous commenter wrote. "Cannibas Cup was a major disappointment. Not a collection of like minded individuals sharing a common interest. Instead, simply a pretext to sell marijuana related wares (bongs, pipes, etc.). Didn't expect nearly that level of commercializism."

The commercialization and the lack of reggae were big topics for the grumblers, and then there were the critics who criticized the critics, as MR SKUNK FUNK did: "Why does every comment i read seem as though the person writing it is either retarded or fully stoned? typo much?"

While I think a lot of these posters raised interesting questions and made valid points—hey, I like reggae—I have to say that the Cup is still a unique event. It's not perfect, and by allowing the people to judge the winners you will always have some level of voter manipulation. Even if you found a way to keep judges from being influenced, you'd still have dissension for the same reason that some people love indicas and some adore sativas: It's almost impossible to come to a consensus. There will always be disagreements and debate because dankness is ephemeral and subjective.

Maybe that's why the industry pros pay attention only to the seed company cups, the sativa and indica categories that are awarded through a traditional, and I would say unbiased, blind tasting.

As Don and Aaron celebrated on stage with their fans and fellow growers, I drifted out to the lobby where I met up with the crew from the Sierras. Slim and Natasha were there, glowing like newlyweds on a honeymoon. They were already making plans to return the following year. The Guru and Cletus appeared to have recovered from their smart shop experience, and Crockett was his usual self, relaxed and friendly, sipping a beer and soaking in the ambiance.

I had been saving a fat spliff of Martian Mean Green— DNA's 2005 Cup winner and 2006 *High Times* Strain of the Year—that Jon at Grey Area had given me earlier that day. So we lit it up and passed it around.

It seemed like a perfect way to end the Cup.

Chapter Twenty-one

Heart of Dankness

The next morning I found myself at Schiphol, drinking a coffee in the terminal mezzanine and waiting for my flight. There was the usual throng of travelers, their expressions ranging from panic to boredom and back, the low rumble of Rollaboard luggage, and the ceaseless sound track of announcements in a smorgasbord of languages.

A British family, sunburned the color of freshly boiled lobsters, wandered through the cafeteria looking for something reassuringly fried to eat. A businessman from Africa sat at a nearby table and yammered into his cell phone. He wasn't happy about something. I saw a couple of Cannabis Cup attendees stroll past. They were both wearing Barney's T-shirts and reggae-striped knit caps and carrying a shopping bag that said "Museum of Hash, Marijuana and Hemp" in bold letters. *They'll have fun going through customs.*

I was booked on the same flight as Crockett and his crew so I wasn't surprised to see the Guru wandering around the cafeteria. He joined me at my table, flopping into the chair like he'd just run a marathon. It turned out he had run a kind of marathon. After the cups had been awarded, after we'd shared the Martian Mean Green, he and Cletus had returned to their rented apartment, dumped all the hash and weed they'd acquired during the past week into one big pile, and spent the night smoking it.

Now the Guru looked a little like a vampire with a hangover. He'd also picked up a nasty cough that's sometimes called "Cup lung." He shrugged.

"We didn't want to waste it."

He heaved a weary sigh, hacked a few times, and then looked at me.

"So? Did you figure out what 'dank' means?"

I once got a fortune cookie that said, "Everything has its beauty but not everyone sees it," which turned out to be a quote from the Chinese philosopher Confucius and a pretty good description of what I'd been trying to figure out. Is dankness the ultimate expression of a plant's potential? Is it a situation? A political stance? A lifestyle? Or does it sometimes manifest as all of the above?

I had been in Amsterdam the previous spring, just in time for Liberation Day, a national holiday in Holland that celebrates the end of the Nazi occupation. The weather leading up to the holiday had been miserable: cold and drizzly and unrelentingly gray. But on May 5, Liberation Day, the skies broke, the sun beamed down, and Amsterdam came out to party.

I had visited Aaron at the DNA Genetics storefront, spending the afternoon chatting and sampling some dry-sift Sleestak resin, so I had one of those clear sativa highs going when I walked out into the streets of the city. I was shocked by the change. For days Amsterdam had seemed underpopulated; people had abandoned the streets for the warmth and dryness of their homes. Now thousands of people were out, riding bicycles, waving the tri-colored Dutch flag, and drinking beer in cafés that had moved nearly every table out onto the street. The canals were gridlocked with boats, and the boats were jammed with people drinking beer, eating cheese sandwiches, and waving flags.

At first I didn't realize it was a holiday. I just thought that was what Amsterdammers did when the sun came out. They went nuts and frolicked in the streets. I suppose the flags flying everywhere should've been a tip-off.

But why not celebrate the weather? It was one of those glorious spring days; the sky was an intense blue; the light was golden and clear. The flower stands were electric with bundles of fresh-cut tulips and the wisteria vines that snaked up the fronts of people's homes were exploding in lavender and purple like the best fireworks ever.

When I got to the little alley where I was staying, I noticed a commotion at the end of the block. Curious, or maybe just happy to stay outside longer, I went to check it out.

The block ended at the Amstel River, close to the Magere Brug, or "Skinny Bridge," and when I got to the end of the street I saw that there were thousands of people packed up against the railing, looking out at the water. I shoved my way between a couple of dudes drinking bottles of Heineken, creating a gap so I could see what was happening.

Hundreds of boats, large and small, clogged the Amstel, and a floating platform bobbed in the middle of the river. A large stage had been set up directly across from us, near the Carré Theater. Two projection screens, the kind you find at baseball stadiums, showed an orchestra seated on the stage, then the camera switched and I could see Queen Beatrix and her family sitting in chairs on the floating platform.

Suddenly everyone was clapping. The conductor came out and took a bow and then ushered a couple of singers onto the stage.

The crowd quieted, almost like it was a moment of silence, and then the music started. Softly, almost a whisper, the orchestra began, allowing the sound to build and grow, slowly swelling until it was so beautiful that I gasped. The strings gently introduced the melody, and I felt like I knew this music. I was sure I did, even though I don't know much about classical music. One of the opera singers, a woman, began singing in a heartbreakingly pure voice. "Just a perfect day, drink Sangria in the park . . ."

It was "Perfect Day" by Lou Reed. It's one of my favorite songs, and I was swept up in the music. And then I was struck by the fact that the Royal Dutch Orchestra was playing a Lou Reed song for the queen of Holland. If that wasn't bizarre enough, they were playing Lou Reed for the queen on their national holiday to celebrate their liberation from oppression.

The music expanded and the singer's voice filled with emotion, conveying all the simplicity and love and sincerity expressed by the song. People around me were wiping tears from their eyes. I was wiping tears from my eyes.

I realized that the song itself could not have been more fitting for the day we were all experiencing.

And for a brief moment I fully understood the meaning of "dank."

I'd like to give special thanks to David L. Ulin for his early reads, insights, and friendship.

I couldn't have written this book without the people who generously gave me their time and answered my (sometimes stupid) questions with kindness and good humor: Aaron and Don from DNA Genetics, Franco from Green House, Jon Foster, the crew of amazingly talented growers in the Sierras, Swerve, Joop Hazenberg, Michael Backes, and Debby Goldsberry.

The intelligence and enthusiasm of Mary Evans, Charles Conrad, Doug Pepper, and Brian Lipson made this a better book and made writing this book more fun than I'd ever expected. I'd like to thank Miriam Chotiner-Gardner, Jenna Ciongoli, Philip Rappaport, and Hallie Falquet for helping guide the manuscript through production, Jonathan Lazzara for marketing savvy, Min Lee for her legal eye, and an extra big thank-you goes to Michelle Daniel for a superlative copyedit.

Big ups to Tod Goldberg, Elizabeth Crane, Mary Otis, Carolyn Kellogg, David Liss, Charles Bock, Craig Caudle, David Suderman, and Matthew Zapruder for their encouragement and friendship, and MacKenzie Smith and Dr. Steven Wegmann for the mojitos and math.

And, finally, I'd like to thank my family, Diana Faust, Olivia Smith, and Jules Smith, for not thinking it all too ridiculous.

Booth, Martin. *Cannabis: A History.* New York: Picador, 2005.

Buruma, Ian. *Murder in Amsterdam: Liberal Europe, Islam, and the Limits of Tolerance.* New York: Penguin, 2006.

Danko, Danny. *The Official "High Times" Field Guide to Marijuana Strains.* New York: High Times Books, 2011.

Donahue, Heather. *Growgirl.* New York: Gotham, 2012.

Herer, Jack. *The Emperor Wears No Clothes.* Van Nuys, Calif.: Ah Ha Publishing, 1985.

Holland, Julie, M.D. *The Pot Book: A Complete Guide to Cannabis.* Rochester, Vt.: Park Street Press, 2010.

King, Jason. *The Cannabible Collection.* Berkeley: 10 Speed Press, 2006.

Mak, Geert. *Amsterdam: A Brief Life of the City.* London: Vintage Digital, 2010.

Marks, Howard. *Mr. Nice.* London: Vintage Books, 1998.

Pollan, Michael. *The Botany of Desire: A Plant's-Eye View of the World.* New York: Random House, 2001.

Rosenthal, Ed. *The Big Book of Buds.* Vols. 3, 4. Oakland: Quick Trading Company, 2007, 2010.

Rosenthal, Ed, Subcool, and Hera Lee. *Dank 2.0: The Quest for the Very Best Marijuana Continues.* Rev. ed. Oakland: Quick American Archives, 2011.

© *Martin Rusch*

MARK HASKELL SMITH is the author of four novels: *Moist,
Delicious, Salty,* and *Baked,* and has written for film and television. A contributor to the *Los Angeles Times* and a contribut-
ing editor to the *Los Angeles Review of Books,* Smith is assistant
professor in the MFA program for Writing and Writing for
the Performing Arts at the University of California, Riverside,
Palm Desert Graduate Center. He lives in Los Angeles.